Funding and Implementing Universal Access:
Innovation and Experience from Uganda

.

Funding and Implementing Universal Access
Innovation and Experience from Uganda

Uganda Communications Commission

FOUNTAIN PUBLISHERS
Kampala

UGANDA COMMUNICATIONS COMMISSION
Kampala

INTERNATIONAL DEVELOPMENT RESEARCH CENTRE
Ottawa • Cairo • Dakar • Montevideo • Nairobi • New Delhi • Singapore

Fountain Publishers Ltd
P. O. Box 488
Kampala
E-mail: fountain@starcom.co.ug
Web site: www.fountainpublishers.co.ug
ISBN 9970-02-518 X

International Development Research Centre
P. O. Box 8500, Ottawa, ON, Canada K1G 3H9
E-mail: info@idrc.ca
Web site: www.idrc.ca
ISBN 1-55250-188-4 (e-book)

Layout and cover design: Robert Asaph Sempagala-Mpagi

Cataloguing-in-Publication Data

Uganda Communications Commission
Funding and Implementing Universal Access: Innovation and Experience from Uganda.
Kampala; Fountain Publishers, 2005
__ p; i13, 13 tables, 4 figures, 3 boxes __ cm.
ISBN 1-55250-188-4 (e-book), ISBN 9970-02-518 X

302.2676 1

Contents

Abbreviations and Acronyms

CPP   Calling Party Pay

GDP   Gross domestic product

ICT   Information and communications technology

ISP   Internet service provider

IT   Information technology

IXP   Internet exchange point

MTN   Mobile Telephone Network

NGO   Non-governmental organisation

POP   Point of presence

RCDF   Rural Communications Development Fund

RFPQ   Request for pre-qualification

SMS   Short message service

UA   Universal access

UAF   Universal Access Fund

UCC   Uganda Communications Commission

US   Universal service

USL   Universal service levy

UTL   Uganda Telecom Limited

WTO   World Trade Organisation

Acknowledgements

The Uganda Communications Commission would like to acknowledge with thanks the financing support it has received from Canada's International Development Research Centre (IDRC) for the publication of this guide and also for previous financing support for a research study on policies and strategies for rural communications in Uganda, and for the development of the Operating Manual for the Universal Access Fund. We are in particular grateful to Dr Constance Freeman, Ms Edith Adera and Gladys Githaiga for the efforts they made to ensure the manual's usefulness. UCC is also indebted to Mr Andrew Dymond and Sonja Oestmann of the Intelcon Research Consultancy Ltd for their previous work with UCC, and for their assistance in developing this guide. Over several years, Intelecon was involved in developing the UCC's universal access policy, the preparation of the RCDF Manual of Operating Procedures, and currently in preparing the first round of competitive auctions for rural telephony licences and Internet points of presence.

Foreword

Uganda has made significant progress toward stabilizing the macroeconomic framework and creating an enabling environment for investment and private sector participation in the economy. This progress is manifested in a stable and low inflation rate and a GDP growth rate of over 5 per cent for the last five years (2000–2004).

In the telecommunications sector, teledensity has grown by a factor of about 14 following the liberalisation and introduction of competition in the sector after the enactment of the Uganda Communications Act of 1997. However, most of the growth in the sector has occurred in urban areas.

Notwithstanding the foregoing, the government recognised very early on in the reform process that communications and poverty reduction are positively linked and that special measures would be required to take communication services to rural areas, where more than 80 per cent of Uganda's population lives.

To facilitate the process of universal access provision, a universal access policy, one of the first such policies in the African continent, was developed. This policy spells out how subsidies could be competitively awarded to private sector operators to enhance their opportunity to provide communications services in rural areas. The principles for the subsidy awards are based not only on the experience of countries where universal access funds have been in use for some time but also on some new and innovative approaches. The Universal Access Policy also covers Internet access provision, information technology (IT) content development, postal services, and training in information and communication technologies (ICTs).

This book describes how Uganda's universal access policy has been developed and how the funds from the country's universal access fund can be accessed. I am hopeful that policy-makers, politicians, and decision-makers, both within and outside Africa, will find this book of use, especially if their governments choose to follow a path similar to that which Uganda has already started.

Hon. John Nasasira
Minister

Ministry of Works, Housing and Communications
Government of Uganda

Preface

Uganda was one of the first countries in Africa to develop a policy on universal access to communications, covering both telephone and information and communications technologies (ICTs). The country's population is over 80 per cent rural, yet Uganda is also the first country in Africa to implement a universal access fund – the Rural Communications Development Fund (RCDF) – operating on the principles emerging internationally as best practice for allocating 'smart subsidies' to private sector companies wishing to serve the universal access market. Several African countries have such funds established in law; almost all are in the very early stages of implementation. In 2004 Uganda completed its first countrywide RCDF tender process and will be well on its way to having public telephones in every one of its 900+ rural sub-counties.

Universal access funds – also called universal service funds or telecommunications development funds – are operational in several Latin American countries as well as in a small number of Asian countries. The earliest funds implemented in the 1990s in Chile, Peru, Colombia and Guatemala were focused on subsidising relatively expensive fixed public telephone lines to very remote and commercially challenging mountainous locations. Over 20,000 remote villages have been reached with a subsidised payphone service. But since that time, two major developments have taken place. First, mobile communications have exploded, due largely to liberalisation and increasing demand. Mobile networks now serve more customers than the fixed networks in the majority of countries around the world. Africa is in the forefront of this development and now has the fastest telecommunications growth of any continent, almost all of which is accounted for by mobile communications.

African consumer demand has proven to be many times the estimates of the so-called experts. The combination of technology and pricing options (demonstrating increasing operator flexibility in creating pricing packages that require fewer calls to be originated to maintain service) also create features that lower-income people find attractive. These features make mobile services a very 'pro-poor' communication medium compared to a traditional fixed service.

The second world development is that universal access – or universal service, as even developing countries sometimes call it – now includes public Internet and information services. The form of access may vary immensely from country to country, but ICTs must now be included in universal access programmes and policies. Here again, the Latin American countries have made some strides, as some universal access funds have supported community telecentres in rural communities. But Africa is not far behind and, here again, Uganda is in the forefront. By mid-2005, every district will have an RCDF contract in place to provide a local Internet point of presence (POP); the majority of POPs will have dedicated high-speed wireless connections that will be available for schools and other leading institutions, cybercafés, telecentres and private customers to make use of within a certain radius of the district centres.

Purpose of this guide

This guide tells Uganda's story in the context of general world developments. It is aimed at assisting regulators and policy-makers, especially in Africa, to develop a successful universal access fund (UAF) strategy and implementation. The Uganda Communications Commission (UCC) wishes to make Uganda's approach and early experience of universal access policy-making and fund operation available. This approach is not expected to be directly transferable to every other country. The theoretical foundation and key principles and features of UAFs may be universal, but each country faces design choices for its own unique characteristics and conditions. However, we are confident that as an African country that has seen the impact of early liberalisation, the mobile explosion, and a range of ICT developments, and that had an early start in the development of a

UAF, Uganda's experience will be useful to regulators and policy-makers in some other African countries.

What the guide contains

Several elements and stages of universal access development can be traced from Uganda's experience but also observed in the programmes of other countries attempting the same task:

O Basic definitions and concepts
O Policy formulation and universal access development
O Estimation of subsidy requirements
O Baseline demand analysis
O Establishment of fund strategy
O Operating principles, organisation and operations
O Subsidy competition, tendering and contract awards
O Tendering, evaluation and award of contracts
O Monitoring and evaluation

This guide describes the basic decisions and activities at each stage. Particular reference is made to Uganda and illustrations are drawn from UCC's experience. But exceptions and experiences from elsewhere are also recognised and referenced.

Eng. Dr A.M.S. Katahoire
Chairman
Uganda Communications Commission

Patrick Masambu
Executive Director
Uganda Communications Commission

Basic Definitions and Concepts

The Universal Access Fund (UAF), named Rural Communications Development Fund (RCDF) in Uganda,[1] is rapidly becoming the best-practice model for special funding to support access to communications by the poor and rural populations in low-income countries. The UAF has a particular role in the context of universal access policy. It is a mechanism to enable a liberalised or liberalising telecommunications sector to motivate and mobilise private sector investment into rural areas by offering 'smart subsidies' and investment incentives to investors. The use of a UAF brings together government policy, intervention and private incentive in a dynamic and mutually beneficial relationship. It thus enables policy-makers to leave behind the previous-generation policy of expecting public sector (or even privately owned) incumbent operators to meet rural service obligations solely through cross-subsidies for a more effective policy. The policy of assigning rural obligations to incumbents has not performed very well in most developing countries, for a number of reasons. The important concepts of this current best-practice approach to rural telecommunications policy are introduced and explained below.

The UAF model importantly fulfills several critical conditions for universal service as defined by the World Trade Organisation (WTO), whose principles on universal service or universal access in liberalising markets set a standard for best practice without actually prescribing specific policies or methods which vary a great deal from country to country, depending on income level and geographical situation.

1

In this regard, the WTO's Reference Paper on Telecommunications Services (http://www.wto.org/english/tratop_e/serv_e/telecom_e/tel23_e.htm) states that:

> Any Member has the right to define the kind of universal service obligation it wishes to maintain. Such obligations will not be regarded as anti-competitive per se, provided they are administered in a transparent, non-discriminatory and competitively neutral manner and are not more burdensome than necessary for the kind of universal service defined by the Member.

The USAF model as described here and practised by Uganda fulfills these conditions. It also harnesses the energies and interest of private telecommunications operators *already active in the country* to compete with one another, as well as with new entrants if they are interested, and to participate in the extension of service to poor rural areas.

Universal access and universal service defined

Universal access (UA) is the policy objective to provide convenient and affordable communications access, on a community basis, through public access facilities such as payphones and telecentres to the whole population. UA may be defined as placing a publicly accessible telephone in every population centre above a certain population size, or placing public phones such as to guarantee that anyone, no matter where he or she lives, need not walk more than a certain distance – for example, 5 kilometres – to reach a phone.

UA is a precursor to universal service (US), which is the objective of making services available individually to every household. This objective is shared by high-income countries with advanced markets. For developing countries aiming at UA in the near- to medium-term while regarding US as the long-term objective is more practical.

Following an initial study focused on rural communications and the development of a strategy for the RCDF, Uganda's universal access policy (UAP) has established the target of providing basic access to communications (voice telephony and a postal outlet) within each sub-county. Basic access to telephony was defined as consisting of at least one public access telephone or payphone if the population is less than 5,000 people and in larger sub-counties one public access telephone per 5,000 people. In the next stage of development, one public phone per 2,500 people would be provided by 2006. Taking into account population density and the typical geographical size of rural sub-counties, this policy has effectively established a minimum geographical target of one public telephone within approximately 5 kilometres of every person in the country.[2] Uganda, as well as many other countries, has also set Internet access targets. These will be covered in later chapters.

How can universal access be achieved?

The purpose of special funding is to address what is commonly recognised as a gap between the financial cost of what government sets as the target for universal access and what a 'liberalised market' will be able to achieve without external inducement. This problem is often referred to as 'the access gap.' However, it is actually helpful to recognise the existence of two separate 'gaps' which must be understood and addressed differently, even though there is some overlapping grey area between the boundaries of the two gaps. We refer to these gaps as the *market efficiency gap*, and the *true access gap*. We use them as a general model, explained below,[3] and compare them in the context of Uganda's specific experience.

Two gaps to address

The gaps can be envisaged in two dimensions – poverty and isolation. Poverty exists in both urban and rural areas. The cost of addressing both poverty and isolation together, as exists in many rural settings, is much higher than that of addressing either one alone. Providing access to the urban poor can be achieved through policies and innovations that are well within the reach of the market, often without much special finance. The main requirement is to allow entrepreneurs the freedom to re-sell and retail telecom services to people who cannot afford their own private communications facilities. On the other hand, reaching some poor *rural* areas may be well beyond the reach of the market. Figure 1.1 illustrates the limitations of the market place.

Figure 1.1 The two gap model

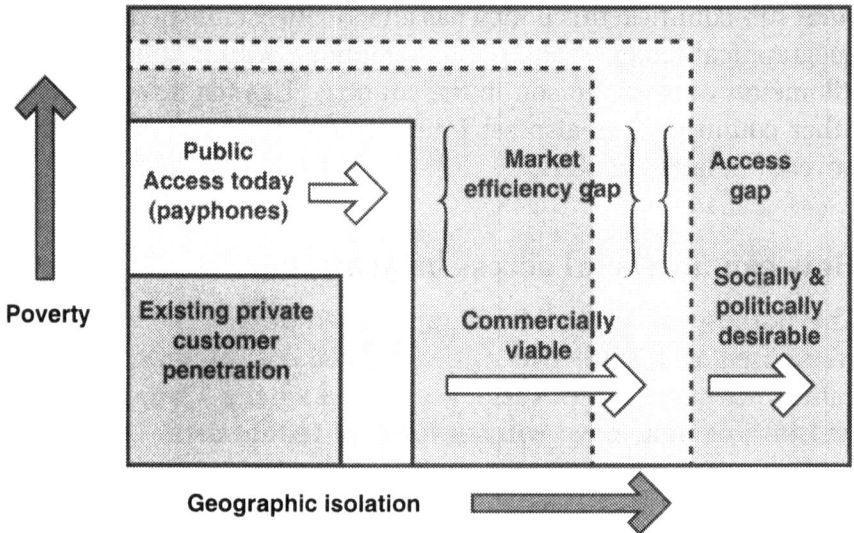

Access to communications goes well beyond the limits of current private residential penetration (represented by the bottom left-hand box in the diagram). The second box is the current access boundary; payphones are traditionally the main means for people without a phone to access the network. Cybercafés perform the same function for the Internet. Also, many private phones, including mobiles, are shared or airtime is re-sold by some mobile owners. This all adds up to a much broader access than pure 'teledensity' implies.

Although this frontier will expand as existing build-out plans and obligatory targets accepted by operators are reached, real-world experience demonstrates that obligations are often only slowly and unwillingly fulfilled. At times, policies and regulations discourage operators or new entrants from reaching or exceeding the obligations, for example, insistence on maintaining inflexible low tariffs that do not allow operators to at least recover costs from obligatory payphones. There is also still the perception that rural areas are loss-making, although affordability is often *greater* than many believe. For example, the large demand of urban residents to call their relatives and friends back in the village (that is, the incoming call market) – amounting to several times the revenue-generating capacity of the rural areas themselves – is largely ignored by conventional telecom economics.

The market efficiency gap

The market efficiency gap is the difference between what markets actually achieve under current conditions and what they could achieve. The gap could be bridged if some regulatory barriers were removed and more market-oriented policies and regulations that create incentives for operators and a level playing field for new entrants were applied. The only questions relate to the market's limits (that is, how far is the commercial reach), and the best ways to implement and sequence more competitive conditions. The outer limits of the market efficiency gap are sometimes uncertain but can be extended by good regulatory policy. Effective market-oriented regulation sets the stage and creates the environment for operators to serve a much broader area and populace and thus close the market efficiency gap. This region can be reached *without subsidies.*

Policy and regulatory measures which can reduce the market efficiency gap include the following:

o   Tariff rebalancing such that services in all areas of the network – international, national long distance, local service rental and calling – are priced to reflect costs and thus to ensure that sector liberalisation does not create distortions (for example, a rush by competitors into areas, such as international calling, that are clearly priced above cost);

o   Tariff flexibility and geographic de-averaging, that is, allowance of a premium, to reflect higher costs in rural areas[4];

o   Market liberalisation – multiple service providers in each market segment, and special opportunities (perhaps with fiscal incentives) for small or niche operators to enter regional and rural territories which are not well served by national operators;

o   Enforcement of separate divisions within the incumbent operator(s) for infrastructure and services – for example, long distance networks, fixed PSTN and mobile subsidiaries – to minimise the opportunity for vertically integrated incumbents to compete and price unfairly;

o   Equal access rules – tariffs and interconnection rights should enforce incumbents to offer competitors, new entrants, niche voice and data operators, to have tariffs and interconnection on the terms offered to related or subsidiary companies;

o   Transparent and cost-based interconnection rules and tariffs that further enable all players – incumbents and new entrants alike – to compete on a level playing field. This should include asymmetric (geographically de-averaged) rates where costs are higher.

Clearly, the range of liberalisation and privatisation measures is wide, and not all are implemented completely in most countries. However, experience shows that they all contribute, in combination, to an efficient market. In Uganda, some of the measures have already been enacted by law or regulatory policy, as explained in the next Chapter. Although its market is still only partially liberalised, Uganda's approach has been remarkably successful in enabling private operators to exceed their initial targets and even to meet some of the objectives that were set as UA targets to need special funding.

The true access gap

Intervention is still required to reach some areas and population groups, even if markets are efficient. These areas and groups will not be served commercially, even with the most attractive, liberal market conditions. They cannot be reached unless additional investments are mobilised through intervention, in the form of subsidies or other special incentives and inducements to service provider entrants.

Policies and strategies to address the real access gap can take different forms, depending on several factors. The depth of target which a country can afford (that is, the population level at which UA targets should be set) must of course take per capita income and geography into account, as these determine the most likely technology and costs. In addition, the stage of existing telecommunications development – network reach, penetration and current liberalisation status – must be considered before the appropriate policy can be defined. However, almost all *access gap* policies today focus on the creation of a UAF, or equivalent, with a combination of (1) competitive tendering between operators to distribute funds for targeted service provision projects or (2) more open-ended competitions for smaller, single-site investments such as telecentres, in which all applicants with acceptable business proposals can secure financial support. The funding sources for the UAFs range from direct government budget contribution (for example, Chile) or proceeds from radio frequency spectrum auctions (Guatemala) to levies from operators active in the sector (Peru, Colombia, India, Malaysia, Russia, Mongolia and elsewhere).[5] The Ugandan model uses the operator levy approach.

Mobile operators in particular are reaching into rural areas through normal commercial means, thus the *market efficiency gap* is being closed naturally through market liberalisation and competition. The mobile expansion is even facilitating better Internet access through the increased presence of digital backbone networks throughout the country. Thus *access gap* policies should focus on the most remote and least viable areas that would otherwise lag far behind in development for many years, rather than target areas of the country that could see commercial service coverage in the short to medium term. At the same time, the actual policies must be carefully matched to the affordability of the country – for example, would the subsidies or investment inducements required to meet a certain UA policy cost the equivalent, say, of 1-2 per cent of total sector revenues per annum, or would they cost more? Some countries have opted for policies that cost as high as 5-6 per cent of revenues.[6] Referring back to the WTO principles quoted earlier, the cost should 'not be more burdensome than necessary' to meet reasonable and appropriate targets and 1-2 per cent appears to be the general consensus in low-income countries. Thus targets and policies generally need to be designed around this guideline. Governments and organisations such as the World Bank are sometimes willing to inject a few million dollars as seed capital to cover the early years before sufficient funds to cover the first years' projects can be built up through operator levies.

Uganda's experience with the gaps

The Ugandan government and UCC has, in effect, addressed the market efficiency gap through several measures that together have led to the UA target being achieved *or exceeded* in the vast majority of the country's 926 rural sub-counties, even before the UAF has held its first subsidy competition. These measures include:

o   Competition in the telecommunications sector was invited even before the incumbent national operator was privatised. This measure created a unique form of market equality between incumbent and new entrant in the marketplace that almost no other developing countries have experienced. Its lesson is a salutory one, as it brought life and vitality to the Ugandan market at a relatively early stage.

o   A stable regulatory regime, with independence strongly guaranteed by law, was created. This regime is understood and trusted by the operators. Interconnection agreements have been fairly balanced and have encouraged network growth.

o   Clear national UA targets, which included a 'serve or lose' clause in the two main operators' licenses with respect to rural sub-counties, were established. By setting clear targets and asking the main operators to declare whether they would meet them by a certain cut-off date, the government and UCC offered both the *threat* and *opportunity* that new entrants would be invited to enter areas in which the main operators did not wish to accept the UA targets. (The UA targets did not become obligations until the existing operators accepted them, at the risk of losing their exclusivity.)

o   In recognising at an early date the potential role of mobile telephony, UCC in 1998 granted a 'technology neutral' licence to a GSM-based operator, as a second national operator, to fulfill its role with fixed or mobile technology, or any combination of the two. The incumbent was allowed to compete country-wide with fixed or mobile technology, as the situation demanded, and a third operator, also a GSM operator, was permitted to offer fixed payphones.

o   By allowing tariff flexibility – operators were allowed to set mobile tariffs for both urban and rural areas – with no requirement to meet fixed line tariffs, operators were encouraged to expand according to their own network economics. They are even allowed to utilise public payphone and 'resale' franchise operators for public access. These operators can charge up to 50 per cent more than prevailing urban rates in rural areas.

o   By publishing its willingness to consider a 'special rural interconnect' asymmetric termination rate for high-cost rural areas, UCC created an additional incentive for operators to participate in the upcoming competitive tender for UA licences.

The combination of these measures has clearly bridged much of the *market efficiency gap* in Uganda, thus reducing the cost of financing UA to that of the *true access gap.* In the case of Uganda, the *true access gap* is represented mostly by challenging sub-counties in very sparsely populated northern and northeastern districts which experience some political upheaval.

In Uganda as elsewhere, Internet service providers are more likely to require subsidies than telephony providers, even to reach some district centres. Internet points of presence will be subsidised for the majority of district centres. Although this measure will not guarantee that Internet access is achievable in most rural sub-counties, it will represent the limit of where the access gap can be closed for the time being.

The Universal Access Fund

The UAF replaces the concept of enforced cross-subsidisation (that is, financing rural phone obligations from internal resources, forced upon monopoly fixed service incumbents, with no external subsidy). Cross-subsidisation was the price exacted by policy-makers from newly privatised operators in exchange for their period of monopoly. Once competition was introduced, it was necessary to find another way of subsidising 'high-cost', 'low-return' or 'loss-making' networks.

UAFs are, in fact, a means for the communications sector to meet the challenge of achieving universal access and country-wide market development from its own resources, and to do so equitably. In most UAF examples to date, money is channelled from license fees, spectrum charges or a special levy (for example, 1-2 per cent) on the revenues of all operators and is set aside to assist those operators willing to serve high-cost, challenging rural areas. Supplying funds to those operators with good motivation and willingness to leverage further investment is a key advantage of the UAF model, even though the whole sector benefits in the long run. The rural market is then developed by those who want to do so, to the benefit of all.

UAFs are usually managed by an entity that is independent of both government and operators. In the majority of cases they are under the auspices of the regulator, though with a separate manager, board of trustees, bank account and reporting procedure. The Ugandan RCDF follows this pattern exactly.

The concept – 'Smart subsidy'

The concept of 'smart subsidy' or 'smart incentive,' is in fact enshrined in UCC's 'Rural Communications Development Policy,' which states that:

> The RCDF shall be used to establish basic communication access, through smart subsidies, to develop rural communications. That is, the RCDF shall be used to encourage commercial suppliers to enter the market but not to create unending dependency on subsidy.

As the policy states, long-term sustainability of the service is the objective, and the subsidy is given only once. It is recognised that a financial enticement is required to invest in difficult areas, but the once-only subsidy must meet whatever capital and operating shortfall is required to carry the investment from loss-making to the point of viability and acceptable rate of return. The operator is expected to meet the roll-out obligations spelled out in the bidding documents and in the license, to provide a stipulated quality of service and to develop the business in a self-sustaining manner, free of further subsidies. The operator is usually not granted exclusivity, only the benefit of being 'first in' to a territory which is unlikely to attract other entrants for the foreseeable future.

In summary, projects that could justify UAF funding and fulfill the 'smart subsidy' condition need to meet the following criteria:

o   The service targets meet government UA objectives as 'socially and politically desirable';

o   The overall service supply profile posesses long-term viability (that is, it is sustainable); but

o   In the short term, the project may be considered financially marginal (that is, offering lower rate of return than that needed by private companies), or risky considering the investment required.

The competitive mechanism – 'reverse auction'

The competitive mechanism for distributing *major* UAF subsidies is usually a 'reverse auction.' The Fund administration studies and estimates the maximum subsidy required to allow an operator to serve a designated area or group of communities, and sets this out as the maximum subsidy available to the winning bidder. Applicants are then invited to bid for a license or service contract to meet specified service obligations in the designated areas, and the license is awarded to the bidder requiring the lowest subsidy, subject to that operator passing a pre-qualification process to ensure that it meets minimum technical and corporate criteria and is capable of providing the required services.

Contributors to the Fund

Each country has a different approach to who should contribute to the UAF, whether just fixed operators, fixed and mobile, ISPs, or even the postal sector. In principle all those likely to benefit from the activities of the Fund should contribute equally in proportion to their revenues. In Uganda, all such communications businesses – telecommunications, post and courier – are required to contribute 1 per cent of their revenues. In turn the Fund supports not just telephony, but all kinds of ICT and document delivery services into rural areas.

Basic Policy Formulation and Universal Access Development

The first stage in universal access development involves formulation of the basic policy. The impetus for a policy will derive from the enabling law(s) for the communications sector.

Legal foundation

In most countries that have enacted a law on communications[7] over the last few years, there are statements which define, among many other matters, the following:

o General objectives for the sector;

o Licence qualifications, terms and conditions for operators to provide various services;

o The role of the ministry responsible for communications, and of the regulator, including the licensing, regulation and monitoring of operators; the regulation of tariffs; the protection of consumers; establishment and supervision of technical standards and service quality; promotion of sector development; and promotion of private sector participation and fair competition;

o The general objective of universal access to communication services, in some cases with specific targets though most often without;

o The establishment of a universal access or universal service fund, with or without specific details as to how the fund will be organised.

In the case of Uganda, the law's statement on universal access sets a clear direction of country-wide access to communications, but does not define actual service targets. Likewise, although the Rural Communications Development Fund (RCDF) was introduced as an instrument of policy to be established and operationalised by the regulator (UCC), the law did not define the parameters of the fund, the amount of money to be accumulated, or the political, administrative and management structure of the fund. Several other recent laws in other countries have followed broadly the same pattern[8].

The government must therefore establish a general telecommunications policy, enabled by law, as well as a specific policy on universal access, before a UAF can become operational. It is also preferable for the legal status and powers of the regulator with respect to the UAF to be set out in the law.

The role of the regulator

As clearly established by the applicable law, in Uganda UCC has the responsibility and power to:

○ improve communications services generally and to ensure equitable distribution of services throughout the country; and

○ establish and administer the Rural Communications Development Fund.

Because this role often falls to the regulator, the country's universal access policy may be prepared under his or her direct supervision or, at a minimum, in concert with or under the supervision of the ministry responsible for communications.

In the case of Uganda, UCC was responsible for the preparation of the universal access policy and its presentation to government, and for drawing up all recommendations for the establishment, administration and management of the RCDF. As a starting point, UCC set itself the following general policy objectives:

(1) Provide access to basic communication services within a reasonable distance to all the people in Uganda;

(2) Ensure effective utilisation of the RCDF to leverage investment in rural communication development;

(3) Promote ICT usage in Uganda.

To develop more specific objectives and targets, and in particular to address the needs of the rural population of Uganda, which comprises more than 80 per cent of the country, and thus to develop a strategic policy framework for rural communications development, UCC engaged the services of an international consulting firm with experience in universal access and rural communications.[9]

Sector review and consultation

To develop a strategic rural policy and a UAF strategy two key activities are required: (1) a sector review and (2) consultation with all stakeholders. A thorough analysis of the telecommunications sector of the country – specifically, the existing infrastructure, market structure and stage of policy development – is needed to tailor the strategy to the available infrastructure and its most realistic 'next steps,' given the country's policy position and market players. For example, implementation of advanced ICT services in regions and areas that do not have the necessary underlying basic infrastructure could have little beneficial impact or possible upgradeability, and at worst become a waste of resources.[10]

The initial sector review and consultation usually takes place, as in Uganda, through a broad study in which the experiences and best practices of other countries with universal access and UAFs are considered along with the status, prospects and regulatory environment of the sector.

Framework for sector review – structural model for ICT

Information and communications technologies encompass a wide range of infrastructure, services, content and applications, with many interdependent parts. Development strategies for the sector, especially in rural areas, need to consider the interdependencies carefully. The diagram in Figure 2.1[11] provides a structure in which to view the underlying telecom and Internet infrastructure; the access/delivery modes; and the demanded content, applications and services.

Within each category, the hierarchy of ICT development is shown to follow more or less the commonly understood market take-up

('S') curve. The logical development is thus from bottom left to top right. However, the *dependency* of the services on the available infrastructure, access modes and services, or of the access modes on the underlying infrastructure, must be understood in a right-to-left direction. The arrows thus indicate the paths of development. In general, each element is dependent on other elements being in place below and to the right.

The diagram is geared to the rural African setting, where *public access* to a basic range of applications and services will be more realistic initially than the achievement of advanced ICT at the household level. Also, within the infrastructure stream (right-hand curve), mobile networks and satellites are shown before a digital backbone. This is the way many African countries' infrastructures are developing, even if only temporarily in that order, whereas in advanced countries the reverse has been true.

Figure 2.1 ICT applications / delivery / infrastructure divide

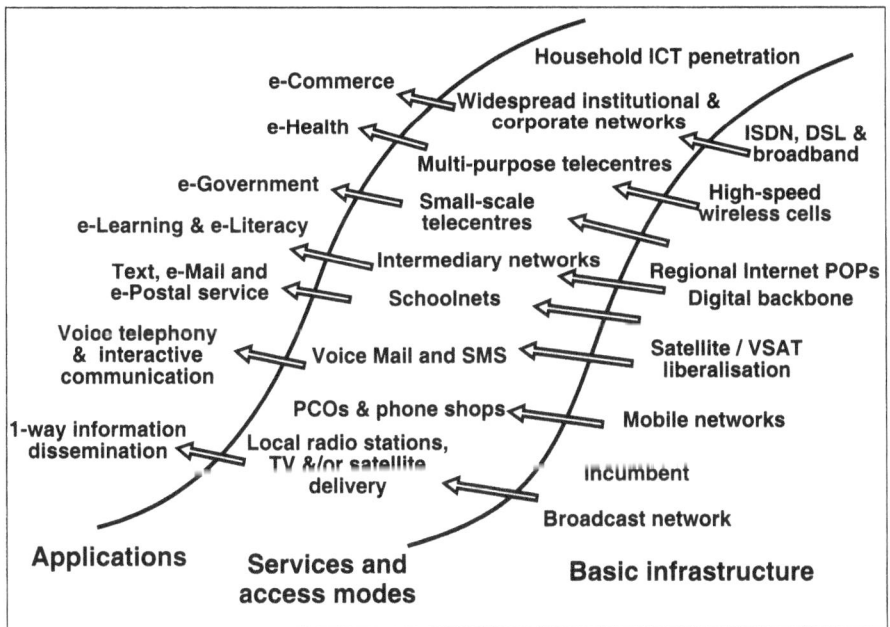

Household ICT penetration

e-Commerce

Widespread institutional & corporate networks

e-Health

ISDN, DSL & broadband

Multi-purpose telecentres

e-Government

Small-scale telecentres

High-speed wireless cells

e-Learning & e-Literacy

Text, e-Mail and e-Postal service

Intermediary networks

Regional Internet POPs

Schoolnets

Digital backbone

Voice telephony & interactive communication

Voice Mail and SMS

Satellite / VSAT liberalisation

PCOs & phone shops

Mobile networks

1-way information dissemination

Local radio stations, TV &/or satellite delivery

Incumbent

Broadcast network

Applications

Services and access modes

Basic infrastructure

However, it is important to recognise, as Uganda has done, that GSM mobile infrastructures are fundamentally digital and tend to create and require the same kind of backbone requirements as a fixed network (that is, they expand from the main cities to regional centres, then to district centres, etc.). Hence there is not necessarily a conflict with mobile network leading fixed network development, so long as planners integrate or make provision for other forms of demand for fixed or semi-fixed services (for example, regional switches and ISP routers, high-speed wireless access systems) at the base station sites. This has been the development path in Uganda, under the universal access policy.

Understanding where the country is on the curve enables identification of existing potential which can be realised rapidly, as well as where support should be targeted to have the most impact. To identify a feasible UA strategy and to develop appropriate targets, the several steps need to be taken and questions answered.

Country characteristics

The geographical, political, economic, social and cultural structure must first be reviewed and summarised – ideally in tabular and graphical format – to present a descriptive array of indicators that reflect the country's characteristics, within the political and administrative structure (for example, in Ugandan terms regions, districts, counties and sub-counties). Population, geographic, topographic, demographic and socio-economic indicators that reflect relative wealth, well-being and poverty should be tabulated and compared regionally. This is required to be able to visualise the context of universal access and the potential role or impact of communications (or the lack thereof) on development.

The result of this initial exercise can be the preparation of a basic desk-based comparative demand study which tabulates socio-economic and infrastructural well-being or poverty and the communications purchasing capacity of the country on a region comparison basis – in the case of Uganda, district by district and sub-county by sub-county. The final output allows a ranking and 'zoning' of the country to be accomplished as an initial starting point. The process is illustrated in Box 2.1.

Box 2.1 Initial data collection and demand analysis

The simplest initial demand estimate can be made in terms of the call-generating and purchasing power of total areas and/or administrative units, as part of the initial country and sector review.

The demand for telephone service has been identified at between 2-5 per cent of a community's total income (that is, per capita income x population, or household income x number of households) no matter the community's level of income. People want and need to communicate and are prepared to pay for it. The benefits they receive for their expenditure are now well known and relate to routine family, business and emergency matters and to time and expense savings (the 'opportunity cost') of alternative means of communications. The actual percentage of disposable income they are willing to spend varies from country to country, but a good rule of thumb for African countries is 2-3 per cent for the desk study at least.[12]

Desk studies start with comprehensive collection and analysis of demographic, geographic and socio-economic statistical data. A tabular 'demand scenario' can thus be created on the basis of total revenue estimates and the distribution of potential private lines, as well as relative Internet requirements, due to the presence of administrative and business or institutional structures, using a range of data:

- population density and distribution
- income levels and distribution
- nature of economic activities and major sources of income
- education levels and enrolment
- health and health infrastructure statistics and indicators
- composite human development indicators (HDIs) if available
- commercial indicators (for example, number of registered bank branches, businesses)
- local governments/administrative offices, other institutions, NGOs, etc.

continued

- proximity to major construction or development activities and programmes
- distance (proximity) to major centres, such as regional city or town, ports, fuel outlets
- physical infrastructure indicators (paved roads, etc.)
- basic infrastructure such as power, public transportation and postal services
- terrain (for example, rugged mountains, hills, savannah, crop lowlands)

The comparative analysis especially provides a description of the country's geographic zones, regions, administrative departments, municipal units and other population centres. The output can provide both descriptive material and a tabular comparative demand analysis, allowing a classification of the country's regions for total market capacity (potential revenue), level of development, and the existence of social and physical infrastructures which create demand for private lines.

Current infrastructure

Together with the basic data analysis, a review of the existing infrastructure that reaches rural areas is required. Knowing the status of rural infrastructure allows us to understand the services and applications it can support as well as the feasible upgrade possibilities, alternatives and potential for leveraging the existing technology for new types of services. For example, countries with rapidly increasing mobile coverage, but no appreciable public access vehicles (PCOs or phone shops) outside towns, may be prime candidates for policies and/or financial programmes focusing on the establishment of phone shops and franchise-like networks. A programme could focus on bringing together partners to develop the public access business, possibly in partnership with the mobile operators. Or if a country has a reasonable digital backbone, facilitating Internet POPs (points of presence) in key regional centres could be the next step, enabling leading schools, intermediary agencies, local government and businesses to get onto the net.

In Uganda, the review of current infrastructure indicated a rapidly growing digital GSM infrastructure reaching most districts of the country and touching many rural areas, creating demand for communications and expectations of further development amongst the population. In addition, the leading operator had established a payphone subsidiary to provide public access to the mobile infrastructure, while numerous individual entrepreneurs were seizing on the opportunity provided by a relatively liberalised environment to establish themselves as roadside or village phone shop operators offering network access to the general public. Furthermore, wherever stable medium or high-speed access to the Internet became available – mainly to date only in the largest district towns – Internet cafés were springing up to offer public email and Internet access.

Policy situation and plans

It is also necessary to thoroughly review the current policy and regulatory framework – including tariff policy, spectrum allocation, licensing and liberalisation. Developing a clear view of likely policy changes in the short- to medium-term is also necessary, as they determine what infrastructure development and service provision could be leveraged for rural ICT projects.

If, for example, the fixed (PSTN) telecom infrastructure is inadequate for Internet deployment beyond main cities, the technical options, short of building a completely new backbone, involve either 'piggy-backing' on the mobile network backbone[13] or use of satellites. The country's regulatory stance on satellite transmission, and specifically VSATs, may be critical to the widespread deployment of Internet POPs beyond the reach of the PSTN or mobile network.

Uganda's small size, combined with its semi-liberalised market status and the presence of digital transmission systems and base station towers in almost all district headquarters, was considered significant, as was the fact that most demand for Internet access in the short to medium term would likely be limited to the area around district towns. This demand could trigger the spread of Internet POPs offering high-speed wireless nodes at the district centres. However, the cost of leasing long-distance bandwidth from the dominant telecommunications operators would hold back deployment by independent ISPs. This issue is not yet resolved by UCC.[14]

Market players and operational stakeholders

It is highly recommended that the existing telecommunications and Internet infrastructure service providers be interviewed about their current experience, costs, revenues, expansion plans (to rural areas) and their view of the rural markets. This often reveals potential opportunities to expand networks and services. 'Market players' also include non-ICT players that have an interest or potential interest in rural areas, be it commercial or development focused. For example, if a country has an active health NGO wanting to reach the rural population, it could use a combination of an FM radio station for information dissemination and call centre and information retrieval services for call-in and 'question and answer' services. This would add to the demand for private and public access telephony and messaging services. Or, if a country has a good banking/credit system, a bank's interest in servicing the demand for financial remittance transfer to rural areas could open up a feasible project to bring e-banking solutions to rural areas.

In some countries, the situation may require concentration mostly on one type of development for some period. In other countries, developments can take place at several levels simultaneously. However, it is essential for the UA development programme to undertake a thorough review of the status quo as well as near-future plans in regard to existing infrastructure, policy and market players. This review would allow feasible UA targets to be created that build on a country's existing strength and potential and that leverage the capacities and synergies of market players and stakeholders already active in the market.

Other stakeholders

In developing a UA strategy, the Ugandan government required that stakeholder workshops be held during the consulting process, for two purposes:

o   To present international experience and best practices relevant to the country for consideration and discussion;

o   To present observations, findings and initial strategic options for a UA strategy for consideration, discussion and feedback by all interested parties, ranging from government officials and market players to rural community-based institutions, community representatives and potential financial partners of the RCDF.

The role played by these workshops in creating buy-in to the UA programme generally and to the specifics of the UAF objectives is immense. For this reason, a total of three full workshops, attended by government officials, operators, donors and non-governmental organisations, were held in Uganda during the UA and RCDF strategy development process.

Outcome: Policy objectives and actions

Primary objectives

The sector review and consultation process places before the government and regulator the basic facts required to formulate a draft UA and rural communications policy containing specific strategic objectives and policy actions, which will give direction and vision, as well as an implementation framework.

In Uganda, the plan was developed in two stages. To bring the overall strategy and RCDF into operation, the sector review and initial stakeholder consultation was followed by a nationally representative baseline demand study (see Chapter 4). The basic UA targets and general strategy set at the conclusion of the study process were to:

o   achieve UA to basic communications – 'Phase 1' was to have public access voice telephony available in 926 rural sub-counties and 'Phase 2' was to achieve a target level of one public access telephone per 5,000 inhabitants in every sub-county by the year 2005.[15]

o   support establishment of a local Internet point of presence (POP) in every district of Uganda.

o   increase the use of information and communications technologies (ICTs), by supporting establishment of at least one 'vanguard' telecentre or ICT project, sponsored by a public or private institution, in every district.

o   ensure effective utilisation of the resources of the RCDF to leverage investment for rural communications development as a viable business, through competitive access to smart subsidies.

o   use *special interconnect* (preferential incoming-call terminating rate) as a means of enhancing rural communication sustainability and minimising subsidy requirements.

Role of existing telecommunications operators

To achieve its UA telephony objective, UCC first requested the two licensed national operators, UTL and MTN Uganda, to declare the sub-counties to which they would be able to provide service (that is, at least one public telephone in the sub-county) by mid-2002. The requirement to meet a minimum service level at the county level was contained in their licences and, in accordance with the licences' specific provisions, the national operators effectively gave up their right of exclusivity in sub-counties that were not included in their declarations (see Table 2.1).

Table 2.1 Operator status report	
Sub-county declared status	**Number**
Served by both operators	356
Served by one operator	416
No coverage (i.e., 'unprotected' by both operators)	154
Total country-wide	**926**

UCC decided to offer the 154 sub-counties that were left 'unprotected' by the national operators for competitive entry and to make them eligible for RCDF subsidies for the purpose of achieving the minimum level of service set by its UA target. In addition, UCC determined to review its definition of basic universal access from time to time, to ensure that its targets develop as the market and communication requirements of the rural populous grows.

Postal services

Uganda's rural communications development policy has several references to postal services. For example, the definition of basic access includes 'public access to where stamps can be bought, and where letters can be posted and collected.' The policy supports pilots for telecom-postal integration in small communications centres and kiosks/shops and requires investigation of models for rural post franchises.

Whereas some other African countries (notably Kenya) have implemented modern postal service targets, including the use of communications facilities and small-town or rural 'telepost' and other public Internet access centres, UCC so far has focused on basic telephony and district centre Internet POPs, as these are the clear priorities. UCC has not initiated basic access for postal services but will review such access, once payphones and Internet POPs have been implemented.

Estimating Subsidy Requirements

A critical outflow of the sector review and desk-based demand study is an initial estimate of the total subsidies required. The following model, which was used after the sector review stage in Uganda, allows for an analysis of the viable and non-viable areas and an approximation of the total subsidies required. The main model, covering the whole country down to the rural sub-county or equivalent level,[16] is for basic telecommunications only, which remains the key need in rural areas. Financial revenue estimates for Internet access are relatively small. However, an Internet model comparing potential revenues and costs for regional district centre POPs should be made.

Initial viability model

The 'maximum allowable subsidy' is the amount an operator will likely require to provide the minimum stipulated level of service to the target area, using least-cost technology.

For every area to receive service through the UAF, the expected public telephone revenue and costs to supply service must be submitted to a 'viability test'[17] which calculates the likely maximum one-time subsidy an operator would require to provide service in a commercially sustainable manner. This exercise should be undertaken by means of a spreadsheet model that summarises certain key 'country characteristics' data for each administrative area (for example, district, sub-county, local government authority) under consideration as a territorial unit for UAF subsidy.

Revenues

The revenue estimate must reflect the following:

○ Population (**p**) of the administrative units under analysis.

○ Local/rural per capita GDP (**i**), estimated as:

- the average of the lower deciles of regional income distribution;

- the proportion of the rural population to the urban population (assuming that the rural population is generally lower income than the urban one);

- an average for the area, from district GDP (for example, using UNDP data).

○ An expenditure or affordability factor (**e**), which is the percentage of income spent on telecommunications. (This factor may range between 2-5 per cent and can be taken initially as the national average of the respective country.[18] (It may be revised later, after a demand survey.)

○ An 'access distance factor,' which is the proportion of rural population (or revenue potential) which will realistically be reached by public pay phones (Estimations for this factor can be made from a consideration of the average travelling distance for users — for example, halfway between the extremity of the area and the central trading centre where the telephone will be located. The relationship is 'inverse,' such as $1/d^x$, where **x** is less than 1 and **d** is the average walking distance.[19])

○ Number of payphones (**n**), based on a population figure to be the desired UA target (for example, 1 per 5,000 people). [A square root relationship is used between achievable revenue and the number of payphones, reflecting the fact that the proportion of revenue captured by the first telephone is larger than subsequent ones. (The combination of (**d**) and (**n**) may together yield a combined achievable fraction of market achievable by public payphones in each area. Experience has shown that this typically ranges between 25 per cent and 75 per cent of the *potential* market as defined by affordability.[20])]

The potential revenue projection from payphones can thus be estimated as follows:

$$R = p \times i \times e \times 1/d^{0.8} \times n^{0.5}$$

Costs

The cost of supplying service should be estimated using the dominant technology utilised by existing operators. This method assumes that existing operators will be interested in expanding into the new areas. This will also normally represent the cost ceiling, because alternative technologies, whether selected by the major operators or new entrants, would generally need to be less expensive than existing technologies to be considered competitive solutions.

The simplest cost estimation methodology calculates the number of GSM or other wireless base stations required to cover the areas under consideration, using a standard 'ideal coverage radius' (for example, 35 km). This radius is modified by a 'terrain factor' which reduces the effective radius that can be covered, due to hills, mountains or other terrain obstacles known to exist in the area. The terrain factor is 100 per cent for ideal radio territory, but almost always less (for example, 25-75 per cent).

Subsidy estimation

The maximum subsidy requirement may be calculated in one of two ways:[21]

(1) Calculate a 10-year cash flow for each service area, showing revenue (with an allowance for demand growth over the cash flow period), capital and operating costs, and estimate the net present value using an assumed cost of capital as the discount rate. A negative NPV represents the maximum subsidy required by the operator to provide service in a commercially viable manner. A positive NPV indicates that no subsidy may be required, although it will be necessary to investigate alternative revenue and cost assumptions to determine the robustness of the calculation.

(2) Make a basic 'benchmark calculation' that assumes a standard pay-back period of capital cost from revenues (for example, 2 or 3 years) to be typical for private high-risk rural telecommunications investments. This calculation will indicate whether, and by how much, the revenues fall short of providing the operator with an acceptable rate of return.

Tables 3.1 and 3.2 illustrate the revenue and cost calculation methods described above, using an aggressive (two-year) pay-back period that would tend to estimate a required subsidy on the high side of the possible range.

Either of the above methodologies provides a first estimate of the maximum subsidy required, though option 2 is simpler. In any case, the methodology should provide an estimate of the *maximum* required subsidy, for two reasons. First, the assumption that public payphones will be the prime instrument for securing potential revenue is pessimistic. In the event an operator uses the same infrastructure to provide private services also (for example, mobile service, data), significantly more revenue could be available from the private customers, even if less of the total market would be attributed to the public phones. In this event, the final revenue may be significantly higher than that calculated and the required subsidy consequently lower. Second, the revenues are based on local (rural) affordability, with no accounting for incoming call revenue from urban customers phoning their rural relatives and friends. Given an asymmetric cost-based interconnect agreement, this revenue could be greater than that from outgoing calls.

Table 3.1 Methodology 1: 10-year cash flow calculation						
Factor - >	Geog. area	Distance (d) outer to central point	Rural	Target no.	Per capita	Potential
Unit/ benchmark ->	Sq. Km	Km		5,000 per phone	$	2.0%
Local Area 1	1,000	18	16,500	4	300	99,000
Local Area 2	1,450	21	11,700	3	270	63,180
Local Area 3	630	14	21,400	5	350	149,800

Table 3.2	Methodology 2: Benchmark calculation					
Factor - >	Distance access factor	Multi-phone multiplier	Base station cost factor	Total capex assuming terrain factor	Capital recovery against benchmark	Subsidy required
Unit/ benchmark ->	$d^{0.8}$	$n^{0.7}$	$300,000 per BTS	33%	0.50 (2 yr paybak)	$
Local Area 1	0.174	2.64	0.26	236,192	84%	48,510
Local Area 2	0.150	2.16	0.38	342,478	37%	189,313
Local Area 3	0.209	3.09	0.16	148,801	201%	0

Initial estimate of UAF-required capital accumulation

From the foregoing calculation a reasonable projection of realistic UA targets country-wide can be made, and the required total subsidy for telephony service provision over the first few years can be estimated. In Uganda, the methodology initially estimated that 150-200 of the 926 sub-counties would require subsidies. As noted above, once the operators were requested to declare which sub-counties they would not serve, the total came to 154.

To develop a total ICT universal access programme, reasonable estimates for subsidisation of Internet POPs, promotion of ICT access (for example, telecentres), training and other items must be added. In the case of Uganda, telephony licence subsidisation has been projected at less than two-thirds of total RCDF expenditures.

For the UAF plan to be realistic, the planned expenditures must be balanced against projected income. In the case of Uganda, two primary sources of income were projected: a 1 per cent revenue levy on all operators in the communications sector and a $5M 'seed finance' grant to be provided by the World Bank.[22]

For the purpose of early consensus-building amongst sector participants and stakeholders generally, an early estimate of projected income and expenditures should be made and discussed, at least in principle.

Table 3.3[23] provides a summary of how the revenue projections for a UAF can be constructed using, as in the Ugandan case, potential revenues from all sector participants, including the postal and courier operators.

Table 3.3   RCDF financial resources

	2000	2001	2002	2003	2004	2005
Total telecom market	84.5	110	158	231	248	264
Total postal and courier market	8	8.4	8.8	9.1	9.4	9.7
Total sector revenue base (US$ millions)	92.5	118.4	166.8	240.1	257.4	273.7
Total resources available (US$ millions)						
Initial endowment	0.40					
UCC budgetary allocation derived from the 1% universal access levy from sector players	0.24*	1.18	1.66	2.40	2.57	2.74
World Bank Rural Transformation Project				4.0	1.0	
Cumulative amount available	0.64	1.82	3.48	9.88	13.45	16.19

* The initial (year 2000) levies were partial only.

Table 3.4[24] provides a revised initial summary of projected expenditures over the period 2002 to 2005, indicating the Fund's capacity to meet expenditures from income, with a safety margin.

Table 3.4 Typical expenditure activities	Investment US$ millions	Proportion %
Public telephony infrastructure for 154 sub-counties	6.00	60
User 'rural packages' (piloting first)	0.25	2.5
Internet POPs and wireless access	1.00	10
Internet exchange point (IXP)	0.10	1
Vanguard telecentre and ICT projects (first seven projects)	0.25	2.5
Vanguard telecentre and ICT projects (one per balance of districts)	1.40	14
Rural post-franchise-support costs	0.50	5
ICT training capacity investment	0.25	2.5
ICT awareness and ICT content creation projects	0.25	2.5
Total	10.0	

Achievements and lessons learned in the first three years (January 2002–December 2004)

The RCDF has fallen well below the original projections of both revenues and disbursements. The start of operator levy collections was delayed by at least two years because of the implementation of tender procedures which were much slower than originally anticipated. Up to two years was also spent in negotiating the terms of the World Bank contribution for the main telephony and Internet POP competitions and in executing the procedure to bring project consultants on board. In the meantime, RCDF did proceed in 2003 with disbursements that could be financed internally from its operator levies. The expenditures or commitments made between February 2003 and December 2004 are summarised in Table 3.5.

Table 3.5 RCDF disbursements 2003–2004	Amount (US$)	% of total
Telephony payphones	66,000	7.9%
Internet POPs (20 districts)	107,000	12.9%
Telecentres (3)	45,000	5.4%
Internet cafés (19)	114,000	13.7%
ICT training sessions (30)	180,000	21.6%
Internet training and connectivity (14)	144,000	17.3%
Content development	176,116	21.2%
Total disbursements	**832,116**	

On the infrastructure front, the RCDF has financed 20 Internet POPs (contract won by incumbent Uganda Telecom), approximately 66 rural payphones (contract won by MTN Uganda), three telecentres and 19 Internet cafés (disbursed to various sponsors in 22 district capitals).

For ICT training, the RCDF has disbursed or committed over $250,000, in excess of its initial target, to 44 training projects or institutions and has spent or committed $176,000 for Internet content development, including the preparation of district web sites.

As might have been expected with first-round experience, the projects have met with a mixed level of success or impact. For example:

o   The initial Internet POP award was won by the incumbent telecom operator, Uganda Telecom. Other potential bidders never expressed interest thus leaving Uganda Telecom to win all the 20 POPs. As a result of the above, under the World Bank-financed ICT component, it has been decided that to avoid creation of monopolies and at the same time promote competition, no more than 10 POPs should be won by one bidder.

o   Some Internet cafe disbursements, which were meant to kick-start the public access market in district centres, were made after other private sector entrepreneurs had already set up cafés. This development has reduced surfing charges in those locations and has at the same time made Internet services affordable.

31

○ Some of the content development work has yet to result in widely used applications. This is perhaps the most complex area of support attempted by the RCDF and will likely become increasingly important and valuable to the market.

The competition for the main public telephony and Internet POPs, to be financed with a contribution from the World Bank Rural Transformation Project, has suffered some delays. A prequalification process was undertaken in the latter half of 2004 and the request for proposal was finally issued October 26, 2004. Bids were received in March 2005.

After tender evaluation, the eventual subsidies to the winning bidders for the public access telephony project, which was tendered in three geographical lots, is expected to be less than US$ 6 million. The subsidy to be granted for the 32 Internet POPs, which were tendered individually with a winning bidder allowance of up to 10 POPs, is slightly above US$ 1 million.

Over one year of the project's delay was caused by the process of harmonising RCDF's tendering procedure and documentation with that demanded by the World Bank Procurement Department. The latter was only in the early stages of developing its requirements for this kind of programme, thus it is hoped that in future cases, delays for the RCDF or other UAFs will be minimal. However, the delay serves to illustrate a potential disadvantage of relying on major institutional support, and opens the question whether the RCDF might have been further ahead in its programme if it had proceeded without such support.

In reality, if the RCDF had organised itself more rapidly, started its levy collection at the commencement of its existence and channelled the levies into the major telephony and Internet POP investment targets which had been developed in year 2001, it could have already established a track record in at least some regions of the country. As it is, the central telephony component of the RCDF programme is now taking place a full four years after the original conceptual design, in a context in which *over half* of the original territorial coverage has been achieved commercially by one or other of the two main operators, before the tender documents had been published.

On the one hand, this development points to a successful bridging of the *market efficiency gap* and illustrates that effective policy, coupled with the existence and publicity associated with a UAF programme (even without major disbursements), can incentivise operators to meet much of a country's rural objectives commercially. By illustrating that some targets can be out of date even before a UA funding programme can be rolled out, the experience points to the necessity to target *real access gap* subsidies to only the most remote, hard-to-reach and unviable areas.

Baseline Demand Analysis

UA programmes are implemented by the private sector and exist as a market, therefore issues of need and application, demand, affordability, willingness to pay, awareness creation, advertising, competition, regulation, and the growth of all of these must be considered in detail. Thus a baseline survey of consumer and community needs, demands and preferences must be carried out.

Demand studies[25] provide information on:

o   Needs for communication of various kinds;

o   Private demand, affordability and user willingness to pay for telephone and other services;

o   Numbers, location preferences and mode of public access points;

o   Technology and service preferences (for example, mobile versus fixed);

o   Knowledge and demand for different value-added and non-voice ICT services.

Various factors – for example, size of country, expected scale of UA programme and whether a pilot project is planned – will determine whether a fully representative national survey or a geographically limited pilot survey is required. In Uganda, a survey which was representative of all four geographic regions and of various types of rural communities – those with and without existing telecommunications coverage – was undertaken.

The rural marketplace

The factors that together create and sustain a market derive fundamentally from the nature of an area's or country's economy and from its socio-economic, demographic and cultural make-up. Geographical factors – in particular, size, topography and population density – also play an important role in defining how viable, commercially interesting or sustainable various telephone and ICT projects can be.

Telecommunications and more advanced (non-voice) ICT may need to be analysed separately or uniquely. However, it is important to consider both voice and non-voice ICT in the data collection stage.

Demand studies start from very cursory, preliminary desk studies (see Box 1 on pages 18–19) based on the data collected at the sector review stage, then proceed to sample or pilot field studies, followed by full-scale representative baseline demand studies. These have several levels of depth and accuracy. Using the different types of demand studies, the market can be profiled roughly in terms of demand for private lines, demand for calls from public phones, total revenue potential, costs, and profitability of the supply business.

Challenges in forecasting demand for non-voice ICTs

Unlike forecasting demand for voice telephony, forecasting demand for non-voice ICTs poses special challenges. To this point, there is no clear-cut demand and revenue stream that can be identified from non-voice communication, even though there could be a significant market in the medium term. Even urban-based ISPs and applications businesses (information, education, or e-commerce) are finding it difficult to estimate their market and meet targets without risk.

The need, demand and application for ICTs in rural areas of most developing countries (but especially rural Africa) can be characterised as follows: basic voice telephony by the majority of the population; and non-voice ICT networks and 'vanguard telecentres' (for example, projects established by leading schools, hospitals, intermediary agencies with active interests, responsibilities and activities in rural areas). Thus local government agencies, social infrastructure and health institutions, schools, NGOs and some business entrepreneurs provide most of the demand, in addition to some private demand via cybercafés in district centres.

With the exception of some of the more 'advanced' or 'vanguard' rural-based agencies or units, the vast majority of the intermediary agency offices with demand for non-voice ICTs today are based in regional or district centres and/or other small towns, but usually not in villages, even though many of their 'clients' are villagers, and villagers may visit their offices often.

Thus the potential user community and demand for more advanced ICT has to be identified in a more consultative fashion than for telephony and often involves more than simple identification of the number of business or administrative units. It includes consideration of how to catalyse partnerships among development, administrative and private agencies to create sustainable and scalable ICT demand. The process typically includes the following:

(1) Through desk research, stakeholder consultation and consensual analysis, identify the type of agencies, institutional units, associations and individual small-micro-medium enterprises (SMMEs) that are the most likely to become users of ICT, based on current activity and interest, development programmes, state of finances and economy, etc.

(2) Identify the distribution and size of these institutions and units (that is, local government, hospitals and clinics, schools, NGOs, business and/or agricultural associations, micro-finance offices, etc.) and profile potential requirements.

(3) Conduct a survey as a 'reality check' to identify the real potential, the ability to absorb investment and use technology, and the required form of partnerships.

(4) Identify the centres – for example, district centres, small towns, – where Internet POPs will best serve the demand.

(5) Amass specific data for ICT deployment coming from other sector players.

(6) Harmonise the data into realistic (and not over-optimistic) demand approximations.

As a result of the above, it may be possible, as it was in the case of Uganda, to make a tabular estimate of the number of short- to medium-term potential Internet users on a district-by-district basis. This estimate enabled an estimate of potential revenues and subsidy for provision of an ISP wireless access system to be made, although the degree of confidence in the estimate is less than that for telephony.

Demand field studies

A demand field study can be designed for telephony, non-voice ICT, or both (as in Uganda). Usually a demand expert designs the survey with the assistance and active collaboration of a local research institution (for example, university social research department or institute). Figure 4.1 shows the various steps of a field study, which should always be based on a prior desk study.

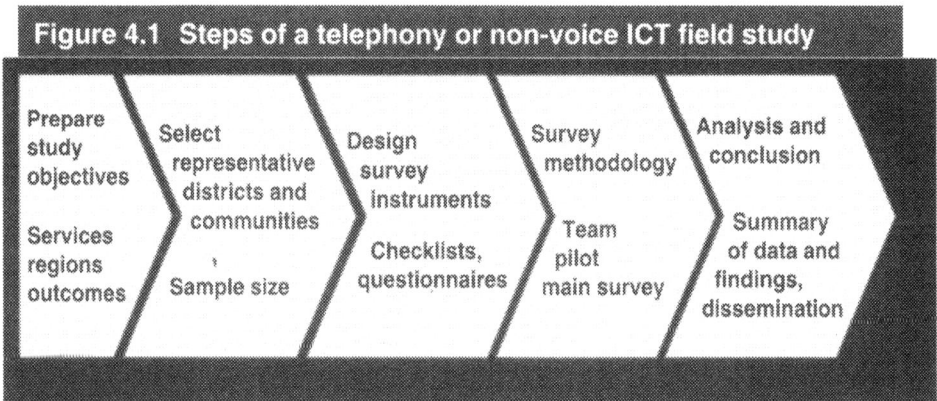

Figure 4.1 Steps of a telephony or non-voice ICT field study

Prepare study objectives	Select representative districts and communities	Design survey instruments	Survey methodology	Analysis and conclusion
Services regions outcomes	Sample size	Checklists, questionnaires	Team pilot main survey	Summary of data and findings, dissemination

Study objectives

The field survey is designed to identify user needs/preferences and ability to pay and to make explicit links to estimates of income and expenditures. Study objectives identify concrete services for investigation, the target groups, and the purposes of the study, and inclusion or non-inclusion of socio-economic impact elements.

Selection of representative districts and communities

A good and careful selection of provinces, districts and communities to study will minimise the need for a large sample. Even if desirable, fully statistically representative field studies are expensive and time consuming, and in most cases their level of accuracy may not be required for the purposes of a UA programme design.

Five questions should guide the selection of a representative sample that can be extrapolated nationally. Does the sample cover:

o   all typical or key regions and districts?

o   typical sizes of villages?

o   both areas with telephone and ICT services and areas without?

o   various economic situations (for example, poor villages and more affluent villages)?

o   different population densities and terrain (for example, remote and sparsely populated, more densely populated, mountainous or plains)?

Survey methodology

An initial pilot survey is recommended to give the survey team a test-run and to refine the methodology and survey instruments. These instruments (questionnaires) should include a mix of the following:

Key informants: Several key informants – ideally in each district surveyed – should be selected and queried about their overall knowledge of the area, their level of expertise and experience in the field of communications and possible use of ICT, and their representation of key priority customer categories. Key informants can provide considerable qualitative information about the area, economy, local village life, etc. They can include district administration officials; representatives of health centres, schools, etc.; community-level leaders / business people / NGOs, etc.; and local representatives of service providers (where relevant).

Household and small business survey: A stratified random sample of a significant number of households and small businesses in each community should be surveyed (minimum of 20). An interviewer-administered questionnaire with both open-ended and non-open-ended questions should be used to solicit responses from both male and female heads of households.

Rapid community market assessment: Community data should be compiled. It should include the number and type of key administrative and social infrastructure institutions, businesses, households, market(s), 'tea houses' and other key social collection points, transportation facilities, etc. which generate demand or locations for public access. Some of these establishments should not be invited to take part in focus group discussions.

Focus group discussions: Several open interviews should be conducted to generate a discussion of some of the key hypotheses and findings of the survey. Participants should be drawn from within the target communities, but should not be individuals who filled in the questionnaires. Special attention should be paid to variations in the socio-economic characteristics of participants and to actual access and utilisation of telecommunication services. Examples of topics discussed include where to locate a public telephone, how far people are willing to travel, how much could they spend, etc.

Box 4.1   Uganda's demand study approach and sample

Uganda's population of 25 million is approximately 80 per cent rural. Apart from Kampala, which is completely urban, there are 58 districts, all with a very high proportion of their population (up to 97 per cent) living in rural communities. The typical structure of 'up-country' Uganda is that each district has a major town (district centre), an average of 3 to 4 counties, 15 to 20 sub-counties, and 90 parishes (each representing several villages). Because of the government's decentralisation policy, the major administrative activities below district centre now take place at the sub-county level. Each sub-county has a 'headquarters' which is usually – though not always – situated adjacent to the main trading centre in the sub-county.

The average rural population density in 2000 was 78 persons per sq. km,[26] but the density ranges from over 300 (for example, Jinja and Mbale) to less than 30 (for example, Moroto 14, Kotido 18, and Kitgum 27).[27]

Large parts of the north have very low population density. The northeast also has the country's lowest literacy and school attendance indicators, and the geographically largest administrative territories. Both the northeast and some northern districts have political upheavals.

In a pilot survey, the lead consultant and research team visited two districts in each of Uganda's four geographic regions: Eastern, Northern, Western and Central. They conducted 43 interviews, primarily in four districts (one in each region), and in 14 sub-counties. Most of the interviews involved several people (for example, a local administrator and friends together, MFI weekly client group, several teachers, parish headman and relatives, groups of traders, focus group meetings). Hence the pilot survey had a broad participation.

The full baseline survey, which was designed after the pilot survey, reflected 640 household and small business interviews at the village level in 32 parishes (20 per parish), spread amongst 16 sub-counties in 8 districts, 2 per region. The selected target communities represented high and low population densities and income levels, places with experience of the telephone and places which were remote from the district centre and telephone coverage.

continued

In addition, the team carried out key informant and focus group discussions in each sub-county. The survey focused on income levels; affordability; interest and willingness to spend on communications; service preferences (for example, manned or unmanned payphones); willingness to travel distances to use a phone; and knowledge of and interest in the Internet, messaging, fax, etc.

The results were summarised in a report to UCC by the research institution conducting the survey and were used by the consultant advising on UA policy and RCDF design to confirm and/or adjust the preliminary desk-based demand projections, UA strategy and subsidy projections.

Development of a questionnaire

Depending on the type of demand requiring investigation, the questionnaire instruments are tailored to the following target groups:

o   Key informants

o   Household users

o   Business users (who or what they are in the rural context should be defined)

o   Institutional users

o   Public phone users

Demand studies should always include a control group of areas that have access to the telephone or other ICT services, if existing. Thus actual existing demand, usage and willingness to pay can be measured. The questionnaire should cover, at a minimum, the areas noted in Table 4.1.

Table 4.1	Areas to be covered in demand surveys
Key information on village	• Other infrastructure (how connected or how remote is village?) • Business structure (retail, agriculture trade, animals, markets, etc.) • Dominant economic activity • Main agricultural crops • Existing institutions in villages (schools, doctor, etc.) • Average household income
	• Age • Family situation • Occupation • Education • Language • Income
Demand Topics	• Current phone availability • Phone usage (frequency, purpose, duration, spending, etc.) • Satisfaction and quality of existing services • Services desired (incl. interest in voice-mail boxes) • Price perceptions, affordability and willingness to pay • Possibility of *incoming calls* • Costs incurred through lack of access to telephone (for example, travel) • Current alternatives: postal services, HF radio, etc.
Socio-economic impact (if required)	• Current communication patterns • Alternative means of commmunication in absence of phone • Cost of the alternative means of communication • Perceived impacts on business and personal affairs • Travel cost savings from use of phone or information services • Other savings or benefits and their valuation

Analysis and summary – outputs

The output of the demand study informs and refines the UA programme design. Demand studies also help to modify the previous estimates of cost, revenue and commercial viability, and the amount of subsidy likely to be required/offered per area or licence.

In addition to providing a comprehensive understanding of affordability, communication patterns and user preferences, specific conclusions from a demand analysis should include the following:

o Indications of the number of public access phones that could be successful per community or neighbourhood – population size to support each payphone;

o Travel distances tolerable for public access;

o Desirable locations for public access phones (for example, trading centres, near village administrative offices, health centres);

o Preference for mode of public phone (for example, manned or unmanned, coin pay, card phone);

o Preferred proprietor for manned phones (for example, shopkeeper, administrative official, school);

o Potential penetration of private service (fixed or mobile) amongst traders, schools, clinics, administrative offices, residences, and so on;

o Preference for technology (for example, fixed versus mobile);

o Knowledge of and interest in voice messaging, 'virtual phone' service, sms text, information services, email, fax, and so on;

o Interest in Internet and potential for telecentre service – which communities have sufficient demand or potential businesses to run a centre;

o Kind of schools, institutions, associations, businesses ready for Internet; and

o Partnerships that could be forged to stimulate sustainable and scalable demand for Internet and ICT services.

In the Ugandan case, the estimation of composite demand (lines/ terminations, traffic and trunk circuit requirements) per community type assisted UCC to design the UA programme parameters, including the telephony licence targets, the most appropriate Internet access strategy and supporting elements of the programme.

Developing the Fund Strategy

The key elements of the financial strategy – that is, how the resources of the UAF will be applied – must involve an exercise in 'strategic zoning' or packaging of the selected targets, regions and territories, to achieve maximum competitive interest as well as reqional equity. It is important to balance several goals such as:

O   attractiveness of the territories to bidders;

O   ensuring no areas are left unserved; and

O   minimising subsidy and costs.

Good strategy will also include adding other incentives for the rural licences such as the right to provide other services or serve additional areas, access to radio spectrum and lower taxes. Also, the tariff and interconnection regime must be considered part of the fund strategy as it influences attractiveness to bidders, the amount of subsidy needed and long-term viability.

Strategic zoning and licence or subsidy packaging

Balancing public and private interests

Establishing a strategy for competitive bidding of subsidies for specific regions or geographical areas necessarily involves UAF planners in reviewing territorial issues. The following are the main issues to consider in strategic zoning and licence packaging:[28]

O   **Sustaining competition by considering operators' strategic objectives:** The fund needs to avoid a situation in which only one or two companies wish to bid. Having several companies interested in bidding will lower the subsidy sought. Companies' appetite for competitive bidding to secure rural licenses under a UAF is related to their strategic objectives. Generally, it is stronger during the initial phases when companies are positioning themselves, but declines when the market has stabilised. However, fund managers need to consider competitive market interests and the strategic objectives of operators in the design and packaging of licence territories. This means that they should assess the apparent commercial interests of the players likely to bid, and package the licence areas in a way that maximises the number of operators that will be interested to bid.

o **The problem of the most marginal localities:** Some licence territories, irrespective of strategic interests, are *much less viable or attractive* than other territories. To avoid absence of interest in these areas, the fund strategy needs to mix attractive and less attractive service areas together and to offer packages that as a whole are attractive although they contain marginal areas. Careful licence packaging is again a crucial issue.

o **Economies of scale balanced with selection options:** On the one side, the strategic zoning of the UA areas into multiple, separately tendered licence territories should give bidders a certain degree of freedom in selecting territories that fit with their network and strategic plans. On the other side, the zoning should not be so fragmented as to diminish economies of scale, and consequently increase the costs of supply and operation.

o **Technology neutrality and opportunities for existing operators:** In general, the strategic zoning should not favour any one operator over others and should be technology neutral, allowing satellite, fixed, mobile or other wireless technologies to be used for cost-effective UA service provision. The only prerequisite should be that operators meet certain technical and service quality specifications which, for rural areas, may be slightly relaxed from specifications expected in urban areas. Whether all existing operators are allowed to provide public services is obviously also dependent on the licensing regime. For example, many countries allow VSAT operators to provide private voice services for corporate clients, but they are precluded

45

from public voice provision. Similarly, mobile operators in many countries are not allowed to provide fixed-wireless or payphone services. The UA strategy is a good opportunity to allow these players to enter the rural market, and thereby increase operators, bidders and optimal use of technology.

o **New entrants versus existing operators:** The UA strategy should be designed to be as attractive to new entrants as to existing operators. However, in some cases this strategy can be tilted towards what is considered desirable for the national sector development. A country with limited competition and few operators could use a UAF process to attract new international entrants to its market. Initially, a rural licence could be given with the promise of a national licence after a period of time. If the number of players in the market is sufficient, the UAF can be focused on enticing existing players to expand their network into rural areas.

Achieving a strategic balance in Uganda

Public telephony

Uganda presented a fairly typical balance of strategic considerations. Tendering the voice telephony packages posed the following issues:

(1) **Uneven packages:** The capital costs and required subsidies for the northern sub-counties are greater than for the remaining regions combined.

(2) **Political stability and security:** The north may also be undesirable because of the security issues: insurgency has been a problem for many years.

To address these main issues, UCC has divided the targeted northern sub-counties into three sections and has included them with the other three regions geographically. Thus Package A (the eastern region package) includes the sub-counties from the north-eastern districts, Package B (the central region package) includes the sub-counties from the central northern districts, and Package C (the western region package) includes the sub-counties from the north-western districts, and as shown in Table 5.1.

Table 5.1 Telephony packages in Uganda			
Package	**A**	**B**	**C**
Region	East and Northeast	Central and Centre North	West and Northwest
Northern districts	Nakapiripirit, Moroto, Kotido	Lira, Apac, Pader	Gulu, Nebbi, Arua, Yumbe, Moyo, Adjumani, Kitgum

The distribution has also been created to balance the number of sub-counties, capital costs and risks associated with each package, as much as possible, as illustrated in Figure 5.1. Thus bidders for each package will be required to accept UA responsibility for some high-cost areas but will be able to obtain the higher subsidies associated with those areas.

During the packaging process it was important to discuss options and intentions with the existing operators, to secure understanding, and to buy-in to the need to include the northern regions. In this way UCC could be confident of participation by the existing operators at least.

Figure 5.1 Balancing of costs and risks in grouping of sub-counties for voice telephony service packages

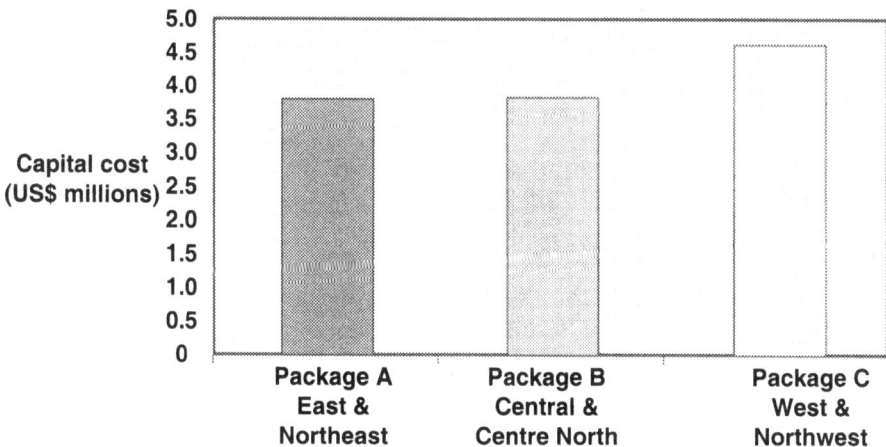

47

Internet POPs

As described earlier, RCDF subsidies are also targeting the district headquarters which have no Internet POP and which operators would not otherwise plan to serve in the near future. The main three issues facing the tendering of the Internet POPs are the following:

o The North may be very unattractive due to security issues.

o Several other districts have low market potential and require much higher subsidies to induce investment.

o The major ISPs in Uganda are closely linked with the main fixed and cellular operators for backhaul from regional areas – for example, Infocom utilises CelTel backbone, UTL uses its own backbone, and Bushnet uses MTN.

To address these issues the following tendering strategy is being pursued:

(1) Bidders will be allowed to select the districts for which they wish to bid so that they can select those which they can serve most cost-effectively given their backhaul situation.

(2) Bidders will be allocated districts for which they are the lowest subsidy bidders, in order of their stated preferences, subject to UCC's guidelines for achieving regional equity. To ensure that the commercially least attractive districts are served, a minimum number of districts will be classed as 'mandatory' for all bidders tendering more than four districts. Bidders will be required to include offers for mandatory districts, according to a formula such as the one shown in Table 5.2.

(3) Bidders will be invited to bid for as many districts as they wish, but no one bidder will be awarded more than 10 POPs unless there are districts with no other bids.

(4) District POPs will be awarded to bidders submitting the lowest subsidy bid, on a POP-by-POP basis, commencing at the top of their list and progressing downwards until they have been awarded 10. Bidders may receive more than 10 if they are the only bidder for some POPs.

Table 5.2 Formula for including offers for mandatory districts	
No of non-mandatory districts tendered by bidder	**Required number of mandatory districts to be tendered**
Up to 4	Not applicable
5-7	2
8-10	4
Above 10	6

This strategy allows the ISPs in Uganda a high degree of freedom to select districts which meet their strategic objectives or capabilities. Smaller ISPs will be able to bid for just two or three POPs if they wish. This selection method should also minimise the required subsidy, as ISPs will bid for districts where they have the best backhaul arrangements. The incentive of higher subsidies attached to POPs in less attractive districts should encourage the larger ISPs to bid for many districts and to be willing to serve many of the less attractive districts along with those that are more attractive.

This tender strategy should help the RCDF to avoid creating a situation in which one ISP is able to establish a monopoly or a dominant situation. Avoiding this situation is more important in the less mature, still emerging ISP market than in the more established public voice market. Uganda has 58 districts, and during the design of the tender strategy only 13 district headquarters had a local Internet POP, 9 of which were implemented on the basis of an earlier tender. Thus the market was in a fairly early stage, and a tender strategy which allows the award of more than 10 POPs to a single ISP would have run the risk of distorting the market and helping one ISP to gain a position of dominance, thereby stifling the development of a healthy competitive ISP sector.

Other bidder incentives

In addition to strategic zoning, there are other means to make rural licences more attractive and less dependent on subsidy, such as the following:

o **Spectrum:** Access to usually expensive spectrum at lower cost could be part of the tendering strategy and could be combined with the maximum subsidy offer. Depending on the situation with already-allocated spectrum, the incentive could be limited to the designated areas or offered nationwide. In other cases, the simple assurance and guarantee that available and requested spectrum will be speadily allocated or that the Fund manager will act as a one-stop shop for spectrum allocation might also be attractive to operators.

o **Licences:** Usually the opportunity to secure a licence is interesting either to new entrants or to existing operators not previously allowed to provide public telephony services and/or use a certain technology. As described earlier, this could apply to mobile operators or to satellite operators. The attractiveness increases if the licence offers operators an opportunity to expand beyond the rural areas or to provide additional services once the UA obligations are met, but this of course depends on country-specific exclusivity periods or other licence guarantees.

o **Taxes:** Import and other taxes can add significantly to the cost of installing and operating infrastructure, even in rural areas where margins are usually slim. Exempting operators from certain taxes for the installation or for a certain period of time can make the bidding more attractive and minimise the subsidy. However, a tax exemption is often more difficult to implement for a UAF as it requires agreement from other ministries and affects their policies and budgets.

In Uganda, the telecommunications market will be fully liberalised by the end of 2005. The additional incentive which new entrants to the rural market could gain is to have established an early presence in the market, while applying in 2005 to become a national operator. However, tendering delays associated with the World Bank's timeline for providing its seed contribution to the RCDF will make this early presence less important as time passes.

Tariffs

Understandably many regulators tend to insist on keeping the tariffs low, especially in rural areas, as they know affordability is lower than in urban areas. However, low tariffs often have an undesired effect. If operators are not allowed to charge somewhat higher tariffs in rural areas, they have no chance of recovering their costs and operating profitably in the long run. The result is low investment in and poor services for rural areas. Whereas affordable service in all areas is the ultimate objective of a universal access policy, it might be beneficial, at least for an interim period, to allow operators to charge, say, 30-50 per cent higher tariffs in rural areas than in urban areas. Furthermore, rural dwellers often develop innovative ways of using the network to their advantage – for example, through sharing phones, extensive use of call-back or 'beeping' their urban contacts who are willing to pay for the calls.

The ongoing viability of a rural operator, after having received an initial subsidy, depends on ability to charge prices that not only cover costs but also allow for a profit margin. Operating costs in rural areas are typically more expensive – for example, maintenance travel, possible need for satellite transmission, more costly E1s, distribution and other marketing activities. Thus a UAF strategy needs to carefully consider tariffs and, in case they are insufficient for rural areas, possibly devise separate tariff regulation for the rural licence bidders.

In Uganda, tariffs are not regulated to single ceiling rates. Instead, the operators have to submit their pricing schemes. The UCC reviews and approves them as appropriate and requires a price cap formula for future changes. The degree of tariff freedom is generally positive for network growth. Effective competition among the three main operators keeps prices in check and reduces them over time, while operators are tending to be innovative in tailoring tariffs for different user groups. For the upcoming rural tender under the UAF, the recommendation for the tariffs within the tendered sub-counties is that they not be higher than 50 per cent of the current average charged in other parts of the country.[29]

Interconnection

Fair interconnect agreements are a regulatory pillar for market liberalisation and are particularly important for the rural market, as good interconnection can limit the size of subsidies required. Ideally, African countries should consider implementing an asymmetric interconnection regime to allow higher termination rates for rural areas, based on a principle of geographically de-averaging. The justification for *asymmetric* termination rates is that rural networks typically cost much more to establish and operate than urban networks, and that urban users would be willing to pay additional tariff rates to cover additional costs to call rural areas.[30]

The cost of rural networks is much higher than that of urban networks: because of lower user density, the length of 'customer loops' or required wireless reach, and the challenging nature of the support structure (for example, lack of reliable power supply). Operational costs such as maintenance and transportation are also costlier. While much has been written – mostly by promoters of wireless technology - to show that costs are generally reducing, the *differential* between rural and urban networks remains such that rural networks may still be several times more costly to establish and operate than urban ones and should therefore have higher termination rates.[31]

There is strong evidence that users in developing countries are generally willing to pay higher charges to cover the cost of higher interconnection. In user surveys, urban users have stated that they would pay extra to be able to call rural communities that have no phone today.[32] In addition, the current worldwide experience of mobile networks using calling party pay (CPP) billing is that customers generally are used to paying for interconnection by paying higher retail tariffs for internet work calling.

These arguments are highly relevant to the developing country and African situation, because universal access must be achieved with limited resources. Thus user pay schemes that would be accepted are important policy tools. Whereas many advanced countries also recognise the cost justification for asymmetric interconnection, they typically choose to maintain geographically averaged tariffs. This is a luxury that only advanced and resource-rich countries can afford.

Developing countries should harness differential interconnection rates as much as possible.

The availability of asymmetric interconnection would provide a better commercial foundation for operators to consider investing in rural networks. Higher interconnect rates for rural networks, would increase operators' interest in taking advantage of pent-up demand for telephone calls from urban centres into low-income rural destinations. A downstream mechanism for meeting this demand could be for operators to share incoming call revenues with public payphone owners, thereby incentivising them to develop the market for incoming urban-to-rural calls and messaging.

Asymmetric interconnection is thus one important mechanism that helps to close the *market efficiency gap* in rural areas and secure ongoing viability after the initial subsidies of a UAF have been exhausted.

In Uganda, UCC is one of the first regulators in Africa that plans to introduce an asymmetric 'special interconnection' regime for rural areas, as well as a rural termination surcharge on calls terminating into public access phones.

Operating Principles, Organisation and Operations

Operating principles

A key success factor in the realisation of universal access through a UAF is the creation of sound operating principles. Stakeholders – such as, rural communities and institutions being served, government ministries being affected, and operators and service providers who contribute funds – need to perceive the fund as credible. Therefore the fund operation needs to incorporate accountability, impartiality and transparency, and efficiency.

Accountability

One of the most important principles of any UAF is proper financial management and accountability. Therefore there must be provisions that safeguard the financial integrity of the UAF, as follows:

o **Separate accounting:** While the UAF administration is typically located within the national regulatory authority, it should be set up as a separate unit with a separate operating budget and accounting for all its costs, for example, administrative and personnel costs. This also requires separate bank accounts with the UAF as the sole authorised user.

o **Rules:** The UAF should have specific rules for its management and staff on procurement, accounting standards and procedures, and accountability regulation.

o **Public annual report:** The UAF management should be required to publish an annual report on its activities, the levies it collected from the operators and service providers, any other funds received, its disbursements and costs.

○ **Independent auditing:** Annually the UAF finances and accounts should be audited by an independent and reputable accounting firm.

Impartiality and transparency

Another key to a successful fund is that the UAF administration has and creates trust in its impartiality within the industry and among other stakeholders, and that it conducts its business justly and fairly, without favouring any particular player. This trust can be ensured through the creation of a special board to oversee the UAF. Whereas the regulator and the UAF administration can set the directives for the UAF goals, management and project implementation, every UAF should have a separate board. Its role is advisory as well as for monitoring. The board should include representation from the communications industry (though not locally active), the consumers, the financial sector, the ministry responsible for communications, and other stakeholders and experts as appropriate. Board members need to declare that they have no financial interests in service providers applying/bidding for subsidies or any other conflict of interests.

Efficiency

The UAF administration needs to be effective and efficient. This requires:

○ Management autonomy, such that decisions can be made and procedures executed without bureaucratic hindrance;

○ Sufficient enforcement and dispute resolution powers to deal with its own industry relationships without the constant need for referral; and

○ Sufficient financial and human resources for timely execution of key tasks.

These operating principles can be realised if the UAF has been established with a mandate, organisational structure, delegated authority, and a manual of operations which provides clear guidelines.

Organisation of the Rural Communications Development Fund

Set-up, roles and responsibilities

The first step in realizing these sound operating principles involves the creation and administrative 'location' of the agency. In Uganda, under the provisions of the 1997 Uganda Communications Act, the RCDF is an instrument of UCC's policy for rural communications development. Therefore the direction, management and implementation of its programme is developed by the UCC. However, the Fund receives part of its finance from the communications industry by means of the levy which UCC collects for the achievement of universal access (the Universal Service Levy). The RCDF resources are aimed at benefiting the general public. Therefore an RCDF board with representation from stakeholders is constituted to monitor and approve its programme and to ensure the transparency of the RCDF.

Dedicated and separate RCDF unit within UCC

UCC appoints a RCDF manager and has allocated staff. The administrative and operating costs of the unit are separate from UCC finances and are financed from RCDF resources. Any other activities UCC undertakes in relation to the RCDF are attributed to the RCDF account. The key role of the Fund manager and his or her unit is to manage the RCDF programme and activities, to implement and administer them, and to manage the finances of the RCDF.

Separate RCDF board

An RCDF board includes representation from the communications industry, consumers and the financial sector. UCC representation constitutes a minority. The function of the RCDF board is to monitor and approve RCDF's programme and activities and to monitor its financial activities.

The RCDF board[33] is made up of the following members:

o   Executive director of UCC

o   Additional UCC commissioner

o   Representative of the communications sector

o   Representative of Uganda Institution of Professional Engineers

o   Representative of the financial sector

o   Representative of a recognised consumers' association

o   Ministry of Communications (*ex officio* representative without voting rights)

To avoid potential conflict of interest, the representatives of the communications industry cannot be owners, shareholders, partners or employees of the licensed communications operators in Uganda.

Respective roles of the RCDF board and the commissioners

As an instrument of UCC's policy for rural communications development, the RCDF board executes UCC strategy. All its operating procedures must be in harmony with UCC's regulatory authority. The Commission has the responsibility to ensure that the activities of the RCDF and the board stay within the overall bounds of UCC policy and regulation.

The RCDF board has approval authority in matters relating to the Fund. However, it is accountable for its actions to the Commission. Moreover, its actions must be consistent with the Rural Communications Development Policy for Uganda,[34] as well as with the procedures set out in the *RCDF Manual of Operating Procedures*.

Without limiting the extent of the board's powers in all matters relating to the RCDF, the board's primary responsibilities include:

o   Approval of the Fund programme;

o   Approval of the operating budget;

o   Approval of all staff appointments other than the Fund manager;

o   Approval of consulting and outsourcing contracts;

o   Approval of all awards of contracts on tendered projects;

o   Maintaining the integrity of the Fund's financial activities in accordance with an established code of conduct;

o   Regular approval of quarterly reports on direct Fund disbursements for small projects; and

o   Monitoring the Fund's quarterly financial reports and annual report.

Fund staff

Depending on the size of the Fund and amount of work to do, Fund staff need to be hired or dedicated. In Uganda, the *RCDF Manual of Operating Procedures* provides for a minimum basic staff of three: the Fund manager, one project officer, and an assistant project officer. Employees of the RCDF are employees of UCC. However, all key staff are dedicated to the work of the RCDF.

Fund manager

The overall management and day-to-day running of the RCDF is the responsibility of a dedicated Fund manager, who has the rank of head of department within UCC. The Fund manager is appointed by the Commission but also answerable to the RCDF board and has the following designated responsibilities:

o   Review of the Fund's investment targets, project plan and budget, in accordance with the UCC Policy and Strategy Paper and the *RCDF Manual of Operating Procedures*;

o   Identification and preparation of rural projects;

o   Preparation and maintenance of the Fund's Manpower Plan: refinement of job descriptions, management of the hiring process, and provision of hiring and manpower recommendations to the board of directors;

o   Preparation, monitoring and control of the Fund's operating budget;

o   Development of project terms of reference, or participation in the assignment of such tasks to project officers or external consultants;

o   Participation in the selection and supervision of consultants to support project implementation;

o   Awareness-raising activities associated with the rural communications programme, advertising and arrangements for public tender notices;

o   Monitoring and following up payment of levies into the Fund by contributors, and acting as the liaison for queries;

o   Liaison with operators involved in rural projects;

o   Initiating funding proposals to replenish the Fund;

o   Budgetary control of Fund and project resources; and

o   Preparation of progress reports to the executive director of UCC and RCDF board.

The Fund may have one or more project officers to manage the operational tasks associated with specific Fund projects. Project officers have the rank of technical officers within UCC. The RCDF may have an assistant project officer (the administrator) who has the rank of an assistant officer within UCC. The RCDF may also request secondment of UCC staff on a part-time or limited period full-time basis for special purposes such as project TOR preparation, tender evaluation or monitoring exercises.

Consultants may be hired by the Fund to undertake and perform a variety of tasks as defined by the Fund manager and approved by the RCDF board. Hiring of consultants and administration of their contracts comes under the responsibility of the Fund manager or of a special sub-committee appointed for the purpose.

Fund finances

All finances are governed in strict accordance with the rules and operating guidelines established by the Fund's annual operating budget. The finances also satisfy all requisite financial policies and regulations of the UCC. The RCDF has its own separate account in a reputable bank with direct foreign currency trading rights. The assets of the Fund may be invested in fixed bank deposits with a bank approved by the UCC, treasury bills and securities of the government, or in other means in accordance with guidelines approved by the UCC.

The Fund manager should ensure that a record of the universal service levy (USL) due to the RCDF is maintained,[35] through the auspices of UCC, and collected from all telecommunications and postal companies. The record should indicate:

O   Verified gross revenues of all companies by year;

O   USL due;

O   Amounts collected; and

O   Amounts outstanding.

Independent auditing of the Fund administrator's bank account, all associated accounts and overall finances are arranged annually by the RCDF's financial institution, in compliance with the requirements of the auditor-general. All payments, withdrawals or other financial transactions relating to the Fund administrator are made in accordance with the Fund administrator *Financial and Accounting Regulations Manual.*

The RCDF board prepares and utilises detailed regulations to govern the financial activities of the RCDF. These regulations are binding on the RCDF board and Fund manager. They spell out legal accountability and liability for RCDF finances.

The Fund manager arranges for and presents to the board quarterly statements of the accounts and activities of the RCDF's finances. The Fund manager also prepares an annual report of all the RCDF's activities, receipts and disbursements. Once approved by the board, and by the Commission, the report is made public.

The Operating Manual

Every universal access fund needs a document which lays down the rules, procedures, principles and guidelines of how the fund is administered and operated. This is sometimes called a UAF operating manual. Contents of Uganda's RCDF are shown in Box 6.1. Each country must adapt its UAF operating manual to its situation and particular fund.

Box 6.1   *RCDF Manual of Operating Procedures*

1. Introduction and Objectives

2. The UCC policy on rural communications
The general UCC and RCDF strategy
Basic access and service definitions

3. Overview of candidate projects and sources of finance
Focus for priority projects
Sources of finance available to the RCDF

4. Governance, management and administration of the RCDF
General constitution, role and functions
RCDF board, role of executive director
RCDF Staff
Project outsourcing, consultants to the RCDF
Finances
Annual report

5. The Fund programme and project selection criteria
The Fund programme
General disbursement priorities
Prioritisation of projects and locations
Subsidy strategy and calculation methodology

continued

---

6. Tendering projects and selection of recipients

General

Pilot funding procedures

Level 1 tendering procedure – Open tender for public telephony, total subsidy above $100,000

Level 2 tendering procedure – Open tender, telephony or Internet, total subsidy below $100,000

Level 3 tendering procedure – Outsourcing for disbursement of small-scale amounts

Level 4 tendering procedure - Open tender for support of district-level ICT projects

Level 5 tendering procedure – Direct disbursement for grants of less than $1,000

7. RCDF subsidy disbursement procedures

General

Telephone licence projects

Internet POP, ICT and bulk purchase projects

'Rural package' contracts

8. Role of UCC in rural operator regulation

General licence conditions

Special rural interconnect

Numbering plan

Customer tariffs

Radio frequency issues

Community participation in public phone development

Public telephone and telecentre management training

Performance monitoring, penalties and enforcement

The *RCDF Manual of Operating Procedures* was prepared and accepted by UCC in April 2002.

Subsidy Competition, Tendering and Contract Awards

Introduction

Disbursement of funds for all projects supported by a UAF should be through competitive or transparent public award and subject to a set of tendering guidelines. The general guidelines on tendering procedures, contracting and subsidy disbursements should be laid out in the *RCDF Manual of Operating Procedures*. Various categories of competition can be considered, each with its own procedure, depending on the nature and size of the project. The actual procedures can range from international open tenders, with documentation prepared in accordance with international best practice, to small grant disbursements against an application pro forma that simply ensures that certain minimum criteria are met. UCC initially defined five categories for competition or application for RCDF disbursements in its manual of operating procedures.

Level 1 and Level 4 procedures are described below in detail. They provide guidance on other aspects of the tendering and subsidy disbursement programme.

Main programme and pilot procedures

The five levels represent the disbursement categories which make up the RCDF Main Universal Access Programme (see Table 7.1). As shown in Table 3.4. UCC decided ahead of time the approximate prioritisation or distribution of funds into each category. In any UAF, the vast majority of funding should be disbursed under its main programme, and it should be disbursed strictly under the procedure guidelines.

However, in the first years of the RCDF, the UCC initiated a few pilot projects to generate interest in the RCDF programme, to establish or refine its knowledge of the rural markets, or to develop its methodology for the main programme. These included the projects listed in Table 7.2.

Table 7.1	RCDF disbursement levels
Level 1	**Open tender:** The disbursement of RCDF funds or public telephony projects, with potential total subsidy amounts in excess of US$ 100,000 shall be by international open tender.*
Level 2	**Open tender:** The disbursement of RCDF funds for Internet Points-of-Presence and training contracts, with expected subsidy amounts of less than US$ 100,000, are by open tender but with invitations publicised domestically and simplified procedure.
Level 3	**Open tender:** This procedure is for outsourcing contracts to facilitate bulk (outsourcing) of Levels 4 and 5 disbursements. The purpose is for offloading the administrative burden from the RCDF, or for the management or franchising of public telephone or ICT businesses.
Level 4	**Open tender†:** The disbursement of RCDF funds to institutions seeking to establish 'vanguard' ICT and community telecentre projects (that is, schools, colleges, hospitals, associations, NGOs or other). These will normally be by open tender *within the district.* The key evaluation criteria will be a business plan demonstrating contribution in cash or kind, and financial profitability and/or sustainability following start-up contribution.
Level 5	**Direct disbursement:** The disbursement of the RCDF funds to applicants seeking support for one or a small number of investments (5 or less) 'rural packages' to enhance signal reception for public telephony kiosks or telecentres, requiring grants equal to or less than $1,000 each. The main criterion for being considered will be a business plan demonstrating financial profitability and/or sustainability following start-up contribution, for provision of services in rural communities that do not have good services.

* Telephony tenders will typically require a subsidy of much more than US$ 100,000 (for example, over US$ 1 Million) however, any subsidy over US$100,000 is considered major.

† Under certain circumstances, small sized subsidies (less than $15,000) may be disbursed directly through approval of a sound business plan.

Table 7.2 Pilot projects and initial RCDF projects

	Amount US$	Status
Public payphones in some underserved sub-counties	66,000	Operational
Local dial-up Internet POPs in 20 districts	107,000	Operational
Internet cafés with ICT training component in districts with existing Internet access	24,000	Operational
Internet content development – preparation of district web sites	87,102	Operational

The process for pilot projects is determined on a case-by-case basis. In accordance with the *RCDF Manual of Operating Procedures*, the project objectives, size of disbursement and the requirements from participants have been decided individually, at the discretion of the Fund manager and with board approval, but have been limited to a small percentage of the overall programme.

UCC considers the pilot programme to date to have achieved its purposes. In particular, it allowed UCC and the RCDF to gain a significant amount of experience and to generate local interest. In addition, it encouraged ICT market growth during the period leading up to the final approval of a significant World Bank 'seed contribution' of US$ 5 million, which will substantially finance the first full round of UA public access telephony contracts.

In other countries (for example, Nigeria), the first pilot project may be a nearly full-scale first-stage implementation of the main programme. In such a case, it is advisable to engage the services of international consultants and to develop all of the necessary tender, licence and service contract documentation in accordance with international best practices from the beginning, in order to test out and refine the procedures and documentation under full operational and competitive conditions.

Level 1 tendering procedure

General

The RCDF public tenders for UA public telephony service offer a
10-year service agreement and operating licence to provide telephone
and other, non-regulated services in the geographically defined areas
targeted by the RCDF.[36] The tender documents define the obligatory
services, which are tied to receipt of subsidy, the amount of which is
determined by the tender process. In Uganda, each service agreement
will require the winning service provider to provide at least one
public telephone in each sub-county within the operating territory
with a resident population of less than 5,000, and one additional
public telephone for each 5,000 people (that is, two telephones are
required for a population between 5,001 and 10,000, three between
10,001 and 15,000, etc.)

The licences will allow the operator to provide other telephony
services – to businesses, institutions and residences at the operator's
discretion – and to provide leased or end-user data services. This
provision is described in the service agreement.

Bidders are invited to bid for the RCDF subsidy for each licence
area, up to a set maximum, which has been established in accordance
with the general methodology described in Chapter 3 and confirmed
in a final design study at the time of tender document preparation.
The exact maximum subsidy amount, and the disbursement schedule,
is stated in the tender documents.

Requirements for public tender

In general the notification process, now being pursued for the RCDF's
UA public telephony programme, has three elements:

(1) **Public advertisement:** Initial notice of tender was advertised
in at least two national newspapers with a wide circulation and
where necessary, in an international journal with wide coverage.
The notice has also appeared on the UCC web site. The initial
advertisement should invite interested companies to request and
pay for an RFPQ invitation document (see below).

(2) **Request for pre-qualification (RFPQ):** The purpose of the RFPQ is to describe the UA programme, the nature of the specific tendered project(s), the service targets, and the general criteria (corporate, financial and technical) which companies will be required to meet to qualify for the competition. The RFPQ stage allows UCC to identify a shortlist of companies. Between 45 and 60 days lead time is provided at this stage to allow bidders to investigate the opportunity, to prepare their responses and to comply with various corporate and legal pre-qualification requirements.

(3) **Bidding documents:** Successful applicants at the RFPQ stage will be invited to purchase the bidding documents. These documents provide the full technical and operational specifications of the required services as well as the draft legal and commercial documentation (the licence and service agreement). A lead time of up to 60 days will be provided for preparation of the tenders from the date on which the bidding documents are made available to the pre-qualified applicants.

RFPQ stage – Bidder requirements

The RFPQ provides detailed project background information on the UA programme, lists the areas to be tendered, provides a preview of the terms of the competition, and outlines all pre-qualification conditions for potential bidders. The actual RFPQ document has been prepared through an international bid consulting contract and conforms to best practices, and has been vetted by the World Bank,[37] which in the case of the RCDF is providing US$ 5 million of seed finance to ensure that the programme can meet all of its basic short-term subsidy targets in the first round of bidding.

Bidders for Level 1 projects in the RCDF competition must generally meet certain minimum criteria, which *include* the following categories and requirements:

(1) Telecommunications operator or service provider with existing experience of providing access facilities in rural Uganda; or

(2) Operator or service provider with existing experience of providing facilities and telephony services outside Uganda, in Africa or in a developing country elsewhere;

(3) Have an existing network exceeding 5,000 subscribers to voice services in rural areas, or a minimum of 250 public payphones and/or telecentres;

(4) Meet a certain minimum standard of financial soundness and ability to provide finance equal to or greater than the expected subsidy on offer (which can be demonstrated in one of several stipulated ways);

(5) Demonstrate good standing and compliance with existing telecommunications licensing conditions and requirements in the company's country of domicile.

Companies that win a bid will have to be incorporated in Uganda before a licence can be awarded to them. Bidder requirements for the rural telephony licences should be kept to the minimum requirements, to encourage a wide variety of companies to consider competing with the incumbent operators. However, the RFPQ requirements are set high enough to limit the final competition to companies that are judged sufficiently experienced and capable of providing the required level of service, extending and/or adding value to the country's telecommunications infrastructure into unserved rural areas, and sustaining and developing a service commercially to a minimum acceptable technical standard over the long run.

Bidding documents

The bidding documents for the RCDF tender are being prepared under the same international consultancy contract as the RFPQ.[38] They include:

o The primary bidding document which contains the project background and all relevant material (including some material already published in the RFPQ) to describe the terms of the tender competition;

o A UA service agreement outlining all aspects of the service provision and subsidy contract;

o A draft operating licence, which will be required in the event one or more new entrants becomes a successful bidder.[39]

Specific designs and conditions of the UA service agreement

The UA Service Agreement is the main document to set out the detailed requirements of the UA services to be provided under the subsidy. These services include network and quality-of-service requirements; construction and subsidy payment schedule; financial arrangements for performance security; project management, reporting and monitoring; and other contractual conditions.

In the RCDF design, the public access telephones which are the essential UA telephony obligation may be outsourced to shop owners, franchisees, local co-operatives, or stand-alone payphone booths. The public phones must be located in the key trading centres of the targeted rural sub-counties. At least some form of private sponsorship or remunerated 'payphone minding' is preferred, to ensure effective operation and maintenance of the public telephones and to develop (as mandated in the service agreement) local messaging in each targeted community by means of human carrier, voice mail, SMS or other electronic means. The network delivery may be by means of extending existing networks or constructing new ones. The bidders may be existing or new operators, or consortia combining network and service retail partnerships.

The service agreement contains a number of schedules which together summarise all technical and service requirements, specific localities, payments and security requirements. In the case of the RCDF tender, the following schedules are included and can be taken as typical for UA public telephony service tenders:

o Schedule A – UA services, availability and quality specifications

o Schedule B – UA service provider's tariffs and interconnection payments

o Schedule C - UA network specifications

o Schedule D - Mandatory service areas and designated trading centres (for placement of public access telephones)

o Schedule E – UA network construction and subsidy payment schedule

o Schedule F – Material events and default

o Schedule G – Security for performance

The network requirements, in accordance with best practice for modern rural telecommunications, are designed to be 'technology neutral,' in the sense that no one solution – for example, GSM mobile, fixed wireless, VSAT, cable or fibre – shall be specified or favoured. However, bidders are required to warrant and demonstrate that the solution offered is operationally and commercially proven elsewhere. In addition, for the purpose of ensuring that subsidies are used to support optimally designed networks and services (requiring as little subsidy as possible), the specification requires that the service provider use technology that is suitable for cost-effectively serving private subscriber demand, over an area of at least 50 per cent of the licensed territories – in addition to the mandated *single point* public access lines – within two years. Moreover the operator will be required to bring the wide-area private subscriber service to market within two years of commencing operation. This requirement will be enforced through a financial performance guarantee.

This provision will tend to favour the use of cellular or other wireless access solutions. UCC's rationale for this requirement is that the subsidies should be used to promote political and economic integration in the target regions. Many people and businesses in the remotest parts of Uganda will need to have access to ICT service technologies – enabling both voice and basic data messaging services – similar to those which currently exist in the rest of the country.

Financial requirements

Bidders for Level 1 tenders need to demonstrate financial and corporate stability sufficient to build and operate the network and services proposed. This can be demonstrated by meeting one or more of several possible requirements such as net worth, cash flow, net earnings and bank borrowing capacity, which together add up to an investment capacity that exceeds the projected maximum subsidy on offer.

UCC also requires bidders to post a performance bond, through a reputable bank, as security for the tendering and subsidy disbursement process. The amount of the bond outstanding reduces as project milestones are met and subsidies are disbursed. The last portion remains in effect until the operator offers private subscriber services in the targeted areas.

Tender evaluation procedure

The tender evaluation criteria are described in the bidding document. Evaluation is undertaken in two stages – technical/operational and financial. First, there will be a public opening of the technical and operational bids; the financial bids will be opened only after evaluation of the technical and operational bids.

Technical evaluation: Within a stipulated period after tender closing, UCC will determine which bidders have met the minimum service obligation requirements and the minimum corporate, technical, operational and quality standards. All bidders will be notified of the full results of this process with a simple yes/no statement covering a number of criteria.

To summarise this process, bidders' tenders will be announced as acceptable or unacceptable based on the following representative, though not necessarily exhaustive, list of criteria (which must have 'pass marks' on all points):

o   Does the bidder meet the minimum corporate qualifications?

o   Has the minimal service level target been offered (for example, 1 public payphone in each sub-county and 1 public telephone per 5,000 people, at the stipulated localities)?

o   Has the required quality of service level been guaranteed?

o   Are the proposed tariffs within the allowed limit for the rural areas set by UCC?

○ Is the technical solution allowable in accordance with the specification and future service requirements?

○ Is the technical solution proven, with at least one reference project with the minimum stipulated number of lines or telephones in service?

Financial evaluation: The financial tenders for subsidy requirement will be opened on a set date, and only after the above notification of technical evaluation has been given. Bidders who did not meet the minimum standard in all basic criteria will not have their financial tenders considered. For each service territory separately, the bidder with the lowest required subsidy of all *qualifying* bidders will be declared the winner. After careful consideration of the option of allowing bidders to make reduced subsidy offers on condition of receiving more than one territory, UCC decided to maintain a simple, single territory evaluation methodology. The reason for this was to discourage, though not to legislate against, the possibility of one bidder winning all three territories.

The *RCDF Manual of Operating Procedures* also allows UCC to *consider* the use of a 'weighted score' evaluation methodology, which would allow consideration of bids with technical and marketing solutions considered to have higher value even if at a higher price. However, since the rules for comparing and weighting the technical/operational bids and financial offers could be complex and controversial, UCC settled for a simple financial (that is, lowest subsidy) selection of qualifying technical/operational bids. This is also judged to be international best practice for UAF tender evaluation at the current time.

Level 2 tendering procedure

The Level 2 procedure is designed for tendering on a domestic-only basis and without the extensive service agreement and licensing requirements of the UA telephony projects. It was designed essentially for the tendering of Internet POPs, in which only ISPs operating in Uganda at the time of bidding will be eligible as bidders.

As in the case of the Level 1 procedure, the tender documents are currently in final preparation under the same international consultancy contract and will contain simplified RFPQ and bidding documents stages. As UCC undertook an early pilot tender for the provision of local-dial Internet POPs at 20 districts, the initial Level 2 competition will involve bids for the balance of districts currently without local POPs. The tender will thus invite bids for the installation and operation of 32 POPs, which will be implemented at district headquarter towns. The POPs will include a high-speed wireless access system to enable users to have a dedicated medium to high-speed Internet service.

The tender will allow an ISP to bid for as many POPs as desired, but contracts will be awarded to the bidder requesting the lowest subsidy, evaluated individually for each POP. No single ISP will be awarded more than 10 POPs unless there are insufficient bidders to apply this restriction.

Level 3 tendering procedure

The Level 3 tendering procedure has not been developed but will be used for investments not covered under the other procedures. Any other RCDF disbursements which are above $25,000 in value will be tendered in an open competitive tender. The tender documents will be drawn up on a case-by-case basis and will be approved by UCC and the RCDF board.

This procedure will facilitate, in particular, the outsourcing of smaller disbursements such as rural public phone wireless extension packages, or other Level 4 and 5 disbursements, to an agency that has the capacity to manage these disbursements efficiently in large numbers. The procedure may also apply to ICT training, capacity building, and other contracts for which intermediate administration may be determined to be beneficial by the board.

Level 4 tendering procedure

UCC has selected a limited number of districts in which to implement the first phase of 'vanguard' telecentre projects. The first phase also functions as a form of pilot for other vanguard ICT projects. The RCDF pilots already undertaken have demonstrated that the funding of commercial Internet cafés will not be necessary or desirable in most districts because they tend to emerge naturally, very soon after a local POP appears. RCDF does not wish to favour one player over another and thus distort a rapidly emerging or changing market. Thus the Level 4 procedure is meant for 'special use' telecentres and ICT projects which, for example, focus on the development of Internet usage by specific economic sectors (such as health, education, media, agriculture), or demonstrate a form of 'public good' in their access programme such as targeting unreached or disadvantaged user groups.

Disbursements for the vanguard telecentre projects will generally be made by open tender among any public or private institutions located within the selected districts. The tender process will involve an advertisement followed by a relatively simple bidding document. The maximum amount of subsidy available, to cover both capital and initial operating cost requirements, is US$ 25,000.

The inclusion of operating costs in the subsidy requirement is limited to the first three years of operation. However, institutions are not necessarily required or encouraged to apply for the full maximum allowable subsidy. Rather they should be guided by sustainability principles for their proposed project.

Key evaluation and selection criteria are the following:

(1) Demonstration of a contribution of at least one-third of total project cost within the first three years of operation (that is, an applicant institution should demonstrate that it will contribute at least US$ 12,500 in value if it applies for a US$ 25,000 subsidy).

(2) Preparation of a business plan which shows viability or sustainability on a continuing basis after the initial three years of operation. The important factor is to demonstrate independence from further RCDF support and the capacity to continue the

ICT project, including depreciation and replacement of obsolete equipment, for at least 10 years.

(3) Demonstration of capacity to implement and manage the telecentre or local area network through adequate and qualified staff and management. Training needs of staff must be clearly identified and can be part of the subsidy requirement.

The application documents and the business plan are required to include the following:

o   Definition of the user community, and an explanation of how the users will be reached and/or sensitised to the opportunity for ICT usage;

o   Clear vision of what the project will address in terms of needs and demand, and the specific ways in which the ICTs will be used by the expected user community, for example, access to educational or professional resources, business/market information, email;

o   How the institution will charge the users (that is, parents) a fee for the sustainability of the ICT project (possibly to cover the operating expenses if no other funding can be found);

o   How the applicant institution(s) will make the facilities available to other users outside its immediate student user community, on a fee-paying basis, that is, to (1) make services available to a wider community and (2) ensure the viability/sustainability of the project as a telecentre;

o   Plans for depreciation and replacement of equipment;

o   Projection in terms of financial cash flow and profitability and/ or educational development for beneficially utilising Internet access;

o   Demonstration of partnerships, where appropriate, to provide ongoing support and resources;

o   Explanation of resources in terms of leadership, management and training required;

o   Item-by-item costing details, along with technical and equipment infrastructure, Internet access costs, and staffing situation;

o   Demonstration of financial viability after the RCDF funding has been used (this viability may be dependent on the amount of support from NGOs, local organisations, parent subscriptions, etc., that is available);

o   The financial plan must show short and long-term financial issues such as: capital costs; ongoing operational costs such as Internet connectivity, staff, office space, marketing (as appropriate), and administration; anticipated revenue from investment, partnerships, donation, RCDF funding, and user fees; in-kind contributions (volunteers, donation of equipment, and space).

Overall, applicants are encouraged to recover costs, to use RCDF resources prudently and efficiently, and to operate economically. The competitive proposals will be requested to be prepared to a standard format, provided by RCDF and published with the bidding documents.

Disbursements will be made in two or more stages: the initial capital investment upon presentation of an invoice for equipment purchase and set-up and the balance at set milestones, to include inauguration (first day of operation), as well as operational stages. The actual payment schedule will depend upon the approved details of each application.

Level 5 tendering procedure

Under the RCDF programme, special rural public telephone packages and other small grant projects may be financed in two ways: (1) matching grants under the direct supervision of RCDF (Level 5 procedure), or (2) grants or loans under bulk disbursement arrangements through the intermediation of an agency employed for the purpose of outsourcing the management of the programme (Level 3 procedure).

Direct supervision by RCDF

In this case, applicants will have the option to get in touch with the relevant RCDF project officer to obtain information about the programme and an application form. The application form will provide an outline for the applicant(s) to describe the location, expected revenues, and costs of the investment and to demonstrate medium-term sustainability.

The applicant shall summarise the items and costs required for installation, up to a set maximum, depending on whether the package includes specific items such as a public access telephone, antenna or power supply unit. The application will have to include technical justification for the investment amount and also demonstrate the applicant's ability to provide the required 50 per cent matching amount of his/her own resources. The applicant will also be required to sign the application, thereby acknowledging the conditions of disbursement.

Bulk disbursement

By this procedure, management of the rural package or other similar programme may be outsourced partially or wholly to an experienced and specialised outside agency through a competitive tender. Such an agency might be able to effectively manage a much larger volume than the RCDF. Also this agency may have special experience in rural business development and micro-finance.

The outside agency would be contracted to identify and/or respond to applicants, assess and approve applications, disburse grants or loans, and verify performance.

At the discretion of RCDF, a revolving fund to provide special-rate micro-credit loans (rather than matching grants) may be used as the disbursement mechanism. Under this scenario, the RCDF grants shall be used to provide the interest rate subsidy on the loan and to cover administrative costs, rather than used as a direct grant to the applicant.

In the case where micro-loans are used as the disbursement mechanism, the competitive tendering procedure will spell out the agreement between RCDF and the contractor with respect to items such as use of funds, relationship to telecommunications service providers, risk assumption, equipment reclaim, and buy-back arrangements.

A pilot project may be used to develop the potential of, and refine the special terms and conditions of this form of contract and programme.

RCDF subsidy disbursement procedures

All RCDF subsidy disbursements – whether in a pilot project or the main programme – are made in tranches, commencing with a down payment upon signature of the service agreement or subsidy contract spelling out agreements and obligations, followed by one or more subsequent payments upon certified completion of the investments and establishment of the project or service.

Payments on *Level 1* tendered contracts shall be made in several tranches, commencing with contract signature and including progress payments up to a final payment after an appropriate period of continuous and ongoing service provision, and are spelled in Schedule E of the UA Service Agreement (Section 7.3.5).

Disbursements on *Levels 2, 3* and *4* tendered projects are made in similar fashion, though agreed in accordance with the reasonable cash-flow requirements of the company or institution carrying out the project. In each case, the final in-service cut-over will be witnessed and/or verified by the RCDF project officer or a representative of RCDF.

Individual 'rural package' or other small-grant disbursements will be made in a more simplified manner, with 50 per cent of the grant amount as down payment upon approval and provision of equipment invoice, and 50 per cent on certification of project inauguration. Approval, invoice and certification forms will be standardised by RCDF.

Monitoring and Evaluation

The importance of monitoring

As the RCDF programme is in its early implementation phase, the monitoring and evaluation processes have not been fully developed. However, UCC has recognised the importance of monitoring and evaluation and integrated them in their concepts and approach. A good monitoring system is important to the success of rural telecommunications and ICT projects. Key reasons for monitoring are the following:

O   If disbursements of funds are tied to milestones, monitoring needs to establish that targets have been met and that the service provision is satisfactory.

O   Monitoring should act as an early warning system and should detect potential abuse and/or difficulties, as well as help to address and rectify problems early in the process.

O   Monitoring can provide feedback to the designers of the next projects and reduce the cost and/or increase the efficiency of post-evaluation studies, thus improving the learning cycle for strategists and planners.

Thus, a monitoring plan should *already be included* in the project planning and development process. This process includes a description of what should be monitored and why, the timeline for monitoring, the resources required for monitoring, the responsibilities of the project owners/participants and how the results of the monitoring process are recorded and used. Both parties usually need to be part of the monitoring effort: the UAF needs to allocate staff and resources, and develop tools and procedures for ongoing monitoring, such as the following:

o an up-to-date and reliable database of all the project's facilities and services, including service roll-out status, service quality and (possibly) usage statistics;   ˙

o performance assessment through a combination of   field visits, basic user and project owner interviews (on a routine basis), and telephone surveys as appropriate;

o a process for filing and resolution of customer or project-owner complaints/difficulties and comments;

o an open communication process (for example, through regular meetings) between funding agency and project owner on the process and progress of the project.

The database and data input process should be managed and updated by the funding agency. The requisite data should be supplied through a system of reports and service checks to maintain relevance. The database could be quite ineffective if not managed by staff dedicated to that role. The staff must be capable of managing the database and identifying exceptions/issues requiring action and resolution.

The inputs can come from external sources through (1) forms and procedures established by the UAF and (2) personnel contracted to collect, analyse and supply data.

There is a cost involved, but monitoring is as important as the initial stakeholder consultation, and the payback in terms of project performance should be significant.

Finally, overall monitoring can be strengthened when monitoring and enforcement mechanisms are built into the regulatory framework and become part of licensing requirements.

In Uganda, the *RCDF Manual of Operating Procedures* states that the Fund manager will require quarterly reports from operators showing network and service status and roll-out statistics, in order to keep a record of total network achievement as well as to monitor operator performance against contract. Typical categories for reporting are shown in Table 8.1. The RCDF Fund manager will make regular reports to the Commission and the RCDF board.

Table 8.1 Operator status report	
	Details
Total number of lines/ accounts in service	• Public phones • Mobile • Fixed business/institutional, residential
Network extent and coverage, total and by district and sub-county	• Sub-county population • Public phone obligation • Public phones in service
Percent completion	• Percent of sub-counties with first public phone • Percent of sub-counties with obligation met • Percent of total network obligation in service
Summary of last month	• Accounts/lines commissioned • Public phones commissioned
Current status of obligation achievement	• Work remaining to complete next milestone • Expected date to reach milestone

To ensure that RCDF-sponsored rural operators provide an acceptable grade of service, to ensure operator compliance with obligations, to determine the need for remedial action, and to create competitive pressure for good performance, RCDF also plans to maintain a database of service quality requirements and operator compliance.

Some of the issues to be monitored and recorded by UCC on behalf of the RCDF are listed in Table 8.2.

Table 8.2	Service quality monitoring
	Activities
Comments and complaints	UCC will publicise the number for customers or service retailers to make enquiries or lodge complaints regarding the service of the operator in their area.
	UCC will keep and file a call record in a simple database or spreadsheet format to facilitate both quantitative and qualitative analysis and response.
Faults and repairs	Operators will be required to organise and publicise their own retailer and customer service centre number; to equip themselves with an operational support and customer care system that can facilitate good response; and to report to RCDF fault statistics, fault diagnosis and time taken to repair.
Call completion	Operators must be able to measure and report call completion statistics and to identify bottlenecks in their own systems and points of interconnection.
Traffic	Incoming and outgoing traffic must be reported.
Billing and retailer support	Operators must be able to deliver to their public telephone retailers complete summaries of per-call outgoing and incoming call times, and monthly summaries, showing calls, average call time, total call time, outgoing and incoming financial amounts owing, and account balance.

Evaluation of rural ICT projects

Again, an evaluation of the various activities of the UAF is important to find out if the strategy and plans have worked, and if they have had the expected impact. This is especially important for Internet-related activities because the Internet market is a less mature and predictable market: the expected usage and applications, and consequently the benefits are still unclear. This information is important for all the stakeholders involved, the government, the Fund, the rural operators and service providers, and the rural population.

A few of the earliest UAFs in Latin America have achieved their initial universal access targets and are currently in the process of developing their second-stage strategy and future plans. A proper evaluation of previous activities is an important prerequisite and provides valuable guidelines for the second-stage strategy. The evaluation of fund programmes should be outsourced through a tender to increase its credibility, and the fund should allocate resources for evaluations.[40]

There are several subjects of evaluation, and each evaluation will look different depending on its specific objectives and focus.

Sustainability

In the case of a basic telecommunications operation, the operating disciplines are well known, and sustainability is reasonably self-evident. Also, the tools of financial analysis, even for rural telecommunications, are well-established. However, sustainability of rural operations extends beyond financial sustainability. This is especially the case in non-basic communications, such as telecentres and other forms of ICT projects. Harris et al. developed a comprehensive model of sustainability that is a valuable tool for evaluating telecentre projects.[41] This model is also a useful way of thinking about sustainability for a variety of other rural initiatives. The elements of sustainability are as follows:

o **Sustaining financial viability:** Financial viability refers to the capacity for income generation that covers ongoing operating costs. Often overlooked are depreciation and equipment replacement costs or upgrades and expansions. Revenues largely come from users paying for services, but this does not mean that other revenue sources such as local in-kind contributions or other donor support cannot be considered.

o **Sustaining staff capability:** It is important to maximise the extent that skilled staff members, or their replacements, continue to stay with the project and that their skills are kept up to date and are properly utilised. This often means paying market-based salaries. Many ICT projects struggle to keep up quality and service delivery because they depend too much on volunteers.

o **Sustaining community acceptance:** If considering the value of the project to the community can support sustainability and also help to protect the infrastructure from vandalism and theft.

Purpose and target

An e-health project will have a purpose different from that of a schoolnet project, e-banking, or a telecommunications network majoring on public access telephones. This type of evaluation addresses the impact of a rural ICT project. The intended or expected impact should be defined and described as specifically and concretely as possible in the planning and development phase.

In general, rural ICT projects are expected to have a positive impact on socio-economic development and poverty eradication. Comparing achievement with expected outcomes in those areas is vital.

An effective monitoring system, leading to good evaluation, will provide feedback on how effective the project targeting or coverage is, that is, how well the project reaches those in need and avoids errors of exclusion. The rationale of implementing a funding subsidy is to distribute a service to those individuals, communities or areas that are not currently covered. If target groups or areas in need are excluded, the mechanism is operating below the ideal.

Replicability and scalability

Many pilot projects remain just that – single experiences. If the fund sponsors pilot projects, it needs to evaluate how replicable the projects are to other parts of the country and how much potential they have to be 'scaled up.'

It is thus important to consider universal as well as specific factors that contribute to the success or failure of the project and, as a matter of routine, to include an analysis of the project's replicability in both the planning and monitoring/evaluation process.

Notes

[1] Or the universal service fund, universal service obligation fund, or the like, elsewhere.

[2] To take a different example which yields a similar result, Russia defines the telephony component of its 'universal service' target as the establishment of a minimum of one public payphone within one hour's walking distance of every inhabited community.

[3] The conceptual framework of the two gaps was developed first in the World Bank Discussion Paper 432, "Telecommunications and Information Services for the Poor: Toward a Strategy for Universal Access," by Juan Navas-Sabater, Andrew Dymond, and Niina Juntunen.

[4] Traditional universal service policy in high-income countries favours geographic averaging in the conviction that everyone should have the same level of service at the same price. This is a luxury that low-income countries cannot necessarily afford and which does not serve the population – even the poor – to the best effect if the country cannot subsidise to the extent that would be necessary to achieve that purpose. Particularly in the case of UA, where people are being offered public payphone service without a requirement to pay a monthly rental, it is far better to motivate operators to provide service at a price that recovers cost and provides a return on investment than to have inferior service or no service at all, since even poor people have a certain level of affordability and derive economic benefit, from making some calls.

[5] In some advanced European countries, such as the United Kingdom, the government or regulator has decided that the responsibility for universal service – that is,, achieving residential penetration and urban-type targets – can be placed upon incumbent operators without the need for a special fund or operator levy, while others (for example, Australia, Canada and the United States – all countries with challenging rural geography) have decided that a universal service fund and operator levy are necessary.

[6] India, Malaysia and the United States are examples, where the government is targeting universal service – that is, high residential penetration – more than public access.

[7] The Uganda Communications Act, 1997, for example.

[8] Nigeria's Law on Telecommunications is a case in which the political and 'management board' and eligible recipient parameters of the 'Universal Services Fund' were defined.

[9] Intelecon Research and Consultancy Ltd. was engaged under the sponsorship of the International Development Research Centre (IDRC).

[10] See also Chapter 4 and 5 in A Rural ICT Toolkit for Africa by Sonja Oestmann and Andrew Dymond, African Connection and InfoDev/World Bank, 2003 (www.infodev.org/symp2003/publications/ruraltoolkit.pdf)

[11] Diagram from the publication in Ref. 10.

[12] Some recent field surveys have indicated expenditure levels of 5-7 per cent in some countries.

[13] Mobile base station towers can serve as the focal points for high-speed fixed wireless centred on Internet POPs (for example, at district centres). Thus institutions and businesses within line-of-sight of a mobile base station could be provided with a direct Internet connection.

[14] The cost of bandwidth charged by incumbent or dominant operators to third-party entities, such as ISPs or end users, compared to the real cost, is a challenge faced by regulators in almost all emerging markets. The operators typically use their own infrastructure to cross-subsidise their own Internet service offerings while charging third parties much higher rates. Strong action is required by regulators to enforce a 'level playing field.'

[15] A Phase 3 for 2005/2006 was later added by UCC to achieve 1 public access phone per 2,500 population.

[16] Ugandan sub-counties have an average population of 25,000 and a geographical area of around 300 km.

[17] See also Chapter 4 in A Rural ICT Toolkit for Africa by Sonja Oestmann and Andrew Dymond, African Connection and infoDev / World Bank, 2003. (www.infodev.org/symp2003/publications/ruraltoolkit.pdf)

[18] The average for Africa is approximately 3 per cent and has been rising steadily as mobile communications becomes the leading mode of service provision.

[19] Although the principle is sound and intuitively rational, there are no major research resources to establish the value of x. The value of x is therefore set subjectively, to achieve results that are supportable from experience.

[20] This estimate is an intuitive and practical means of downscaling the revenue projections from the maximum (affordable) level to a practically achievable level, based on rational factors which have been generally observed through many field visits.

[21] Both require significant assumptions, for example, cost of supply (which can vary a great deal depending on the operator's total traffic capacity assumptions), required rate of return, and pay-back period. Hence, the two methodologies are approximately equivalent, except that the second methodology is easier for a first approximation.

[22] The World Bank contribution has been made under its Rural Electrification Programme.

[23] The figures used in this table are based on forecasts made during the time of initial study and are not actual figures. Actual figures are reported to be below the forecasted figures.

[24] Again, the figures used are based on projections and do not represent actual expenditures.

[25] See also Chapter 5 in A Rural ICT Toolkit for Africa, Sonja Oestmann and Andrew Dymond, African Connection and infoDev / World Bank, 2003. (www.infodev.org/symp2003/publications/ruraltoolkit.pdf)

[26] Uganda's average rural population density is about four times the African average.

[27] The Least dense sub-counties in Karamoja have over 900 sq. km area and just 10,000 population.

[28] See also: <u>Rural Telecommunications Development in a Liberalising Environment: An Update on Universal Access Funds</u>, Andrew Dymond and Sonja Oestmann, World Bank Rapid Response Unit, 2002.

[29] See also: 'The role of sector reform in achieving universal access,' in ITU: <u>Trends in Telecommunications Reform 2003 - Promoting UA to ICTs – Practical Tools for Regulators</u>, Andrew Dymond and Sonja Oestmann.

[30] See also: 'Telecommunications Challenges in Developing Countires - Asymmetric Interconnection for Rural Areas,' Andrew Dymond, World Bank Working Paper 27, 2004.

[31] Historic cost differentials quoted by the ITU are that rural lines are seven times costlier than urban lines. Many examples and graphs can be used to support a similar differential today. Edgardo Sepulveda (McCarthy Tétrault), in a document to CTO and ITU entitled 'Model Universal Services Fund Policies and Procedures,' Part II, June 2002, notes from Cribbett (2002) that 'average lines costs in low-density areas of Australia were found to be between 6 to 10 times the average cost per line in the rest of Australia.' Per-line costs in one of the Chilean rural networks, at $5,000, are clearly at least 6-10 times typical urban costs, and this is reflected in the cost modelling exercise undertaken to establish interconnection rates in that country. Many similar cases in developing countries can be cited. Low-capacity cellular systems will also cost more per subscriber or per user-minute than high-density ones.

[32] 'Policies and Strategies for Rural Communications in Uganda,' March 2001, a report submitted by Intelecon to UCC, contained an extensive user baseline survey that documented, among other things, a high demand for urban-to-rural calling and willingness to pay higher on urban-to-rural calls.

[33] The RCDF Board shall have the right to propose additional representatives from the IT, telecommunications and postal sectors, as the need arises.

[34] Prepared in July 2001 and accepted by Parliament.

[35] The UAF accounting may be carried out by UCC or by a dedicated RCDF accountant, but is likely to be the former.

[36] 154 rural sub-counties were specifically targeted for the placement of public access phones, divided into three approximately equal geographical areas, defined as West, Central and East.

[37] The RFPQ is an 11 page document with three annexes.

[38] At the time of writing the bidding documents are in the final preparation stage.

[39] In the event an existing operator wins one or more of the competitions, a relatively simple addendum to their existing licence would be required, indicating the addition of new UA service obligations in consideration of the subsidy award(s).

[40] The Chile and Peru UAFs have conducted detailed evaluation studies, with the assistance of international consultants.

[41] Harris, R.W., A. Kumar and V. Balaji. 'Sustainable Telecentres? Two Cases from India.' 2002. (http://www.developmentgateway.org/node/133831/sdm/docview?docid=442648)

Index

Contacts

Intelecon Research and Consultancy Ltd
400-1505 West 2nd Avenue
Vancouver, British Columbia
V6H 3Y4 Canada
Phone: (1) 604-7382004
Fax: (1) 604-7385055
Email: contact@inteleconresearch.com

Uganda Communications Commission
12th Floor, Communications House
Plot 1 Colville Street
P. O. Box 7376, Kampala
Phone: (256) 41-339000, 31-339000
Fax: (256) 41-348832
Email: ucc@ucc.co.ug
Website: www.ucc.co.ug

www.ingramcontent.com/pod-product-compliance
Lightning Source LLC
Chambersburg PA
CBHW071138280326
41935CB00010B/1276

* 9 7 8 9 9 9 7 0 0 2 5 1 8 3 *